Thomas Sterry Hunt

**A New Basis for Chemistry**

A chemical philosophy. Third Edition

Thomas Sterry Hunt

**A New Basis for Chemistry**

*A chemical philosophy. Third Edition*

ISBN/EAN: 9783337070168

Printed in Europe, USA, Canada, Australia, Japan

Cover: Foto ©berggeist007 / pixelio.de

More available books at **www.hansebooks.com**

# A NEW BASIS FOR CHEMISTRY:

## A CHEMICAL PHILOSOPHY.

BY

**THOMAS STERRY HUNT, M.A., LL.D., (Cantab.),**
AUTHOR OF "CHEMICAL AND GEOLOGICAL ESSAYS," "MINERAL PHYSIOLOGY AND PHYSIOGRAPHY,"
ETC., ETC.

---

Omnia mensura et numero et pondere disposuisti.
—*Lib Sapientiae, cap.* xi.

---

THIRD EDITION, WITH NEW PREFACE.

SCIENTIFIC PUBLISHING CO.
NEW YORK.
1891.

COPYRIGHT, 1890,
BY SCIENTIFIC PUBLISHING CO.

To

## J. B. STALLO,

CITIZEN—JURIST—PHILOSOPHER,

THIS VOLUME IS GRATEFULLY

DEDICATED BY THE

AUTHOR.

# PREFACE.

It is now more than thirty-eight years since the author, repelled by the contradictions which result from the introduction into chemistry of the atomic hypothesis, and seeking a better foundation than that on which Dalton endeavored to build a theory of the science, began a series of publications, having for their object, in the words employed in 1853, to define the "principles which may serve as the basis of a sound theory of chemistry," and at the same time "to enlarge and simplify the plan of chemical science." In pursuance of this plan to the present time, he has persistently followed the line of argument foreshadowed in 1848, and fully stated in 1853. Advances were made in the succeeding years; but the crucial problem, the solution of which was required to complete the projected philosophy of chemistry, namely, that of the relation of equivalent weight to specific gravity in liquids and solids, remained unresolved until the summer of 1886. With its solution, the author believes that the object so, long ago proposed has been attained, and that we have the elements of a new and simple Chemical Philosophy, which he ventures to designate A New Basis for Chemistry.

Rejecting all hypotheses regarding the constitution of matter, as irrelevant to the study of chemistry when rightly understood, and as pertaining to the wholly distinct realm of dynamics, the author has attempted to define the limit between the two, and to show that the phenomena of solution, fusion, volatilization, liquefaction, solidification, and crystallization,— in a word, all changes of state,— belong to the domain of chemistry, and that, from a right consideration of these, some of the most obscure problems of the science become clear. He has therefore endeavored, in a series of chapters, and, in large part, in language quoted from his earlier writings, to define the principles upon which he believes may one day be built a New Chemistry.

The reader will note that the plan and the

limits of this little volume exclude the discussion of many subjects which would find a place in a systematic treatise on chemical theory. Among these are the consideration of the periodic law in relation to the law of numbers, the question of valency, and the problems of thermo-chemistry, all of which must be reserved for another time and place.

It would be unjust on the part of the author not to express his many obligations to J. B. Stallo, now the American Minister at the Court of Italy, whose suggestive volume on the Philosophy of Nature, in 1848, was, at that early date, a source of inspiration; whose friendship during many years has encouraged the writer in his labors; and whose later volume, on The Concepts and Theories of Modern Physics, in 1882, has served to confirm him in the position which he has long maintained with regard to the philosophy of the sciences. To Stallo, therefore, this book is dedicated.

BOSTON, January 1, 1887.

# PREFACE TO SECOND EDITION.

A YEAR has passed since the publication of the first edition of this volume, and a second is called for. In that interval the author has prepared and read before learned societies, in farther extension of the views therein put forth, four papers, two of which have been published. He has, moreover, detected an inadvertence by which, alike in the first edition and in these published papers, a grave error was introduced into his previous calculation of the integral weight of water, and consequently of all species for which that liquid is the unit of specific gravity. Portions of Chapters IX., X., XI., and XIII. have accordingly been revised, and a supplementary Chapter of seventy-six pages, with an Appendix, has been added, containing the substance of the four papers mentioned. A new Index has also been prepared.

A translation into French of this second edition, by Professor W. Spring of the University of Liège, in Belgium, is in preparation, and will soon appear.

MARCH 15, 1888.

# PREFACE TO THIRD EDITION.

THE prefaces of the first and second editions of this volume leave little to be said on the present occasion. Attention may, however, be called to the fact that a farther consideration of the question of complex inorganic or mineral acids, already noticed in Chapter VI., has led the author to regard the so-called double silicates of alumina and protoxyd bases, as aluminosilicates of these same bases, the alumina and silica together constituting a complex mineral acid; in which, moreover, the alumina may be wholly or in part replaced by the feebly basylous or acidic chromic, ferric, manganic, titanic and boric triad oxyds. With the latter we have borosilicates, and in the case of tourmalines fluoroboraluminosilicates.

This view is set forth more at length in the preface to the second edition of the author's MINERAL PHYSIOLOGY AND PHYSIOGRAPHY, and will, moreover, be found embodied in his recent paper entitled, "*Notes on Valency, Basicity, Complex Acids and Chemical Notation.*"

*Preface.*

in the CHEMICAL NEWS for June 6th, 1890. In that same paper the question of chemical notation, here discussed in the concluding pages of Chapter XIV., is resumed, and an attempt is made by the use of different kinds of type for chemical symbols to attain as far as possible, a monadic notation. All of these points will be discussed more at length in the author's SYSTEMATIC MINERALOGY.

It remains to be said that the second edition of this volume has been translated into French by Professor W. Spring of Liége, and, with an important preface by the translator, was published in 1889 with the title of UN SYSTÈME CHIMIQUE NOUVEAU, in 1889 by Georges Carré, Paris, and Marcel Nierstrasz, Liége.

T. S. H.

NEW YORK, Nov. 1890.

# CONTENTS.

## CHAPTER I.

### INTRODUCTION.

A new theory of chemistry proposed . . . . . . . . . . . 1
List of the author's publications thereon from 1848 to 1886 . . 1
Their partial republication in 1874 and 1886 . . . . . . . 5
Their analysis and the completion of the theory attempted . . 6

## CHAPTER II.

### NATURE OF THE CHEMICAL PROCESS.

Chemistry as distinguished from physics . . . . . . . . . 7
Origin of chemical species by metagenesis and metamorphosis . 8
Metamorphosis by expansion and by condensation, or polymerism, 8
Chemical genealogy; double decomposition . . . . . . . . . 9
Polymerism in sulphur-vapor; in carbon and phosphorus . . . 10
Nature of double decomposition; catalysis . . . . . . . . 13
Genealogy of species; their homogeneousness . . . . . . . 15
Chemical identification and differentiation . . . . . . . 16
Chemical integration and disintegration . . . . . . . . . 17
Crystalline individuality; chemism, dynamics, biotics . . . . 18
Mineralogy and biology; physiology of matter . . . . . . . 18
Classification of natural sciences tabulated . . . . . . . 19
Chemism often confounded with dynamism . . . . . . . . . 20
The activities of matter on different planes . . . . . . . 21
Solution defined as chemical union . . . . . . . . . . . 22

## CHAPTER III.

### GENESIS OF THE CHEMICAL ELEMENTS.

| | |
|---|---:|
| Oken; his Physiophilosophy and scheme of the sciences | 23 |
| Genesis of elements from a primal matter | 23 |
| Cosmogony; stoichiogeny, or the genesis of elements | 24 |
| Studies of Dobereiner, Pettenkofer, and Dumas | 24 |
| Prout's hypothesis of equivalent weights | 25 |
| Brodie's Calculus and his ideal elements | 26 |
| Celestial chemistry; universality of dissociation | 28 |
| Brodie's Ideal Chemistry in relation to dissociation | 29 |
| Clarke on stellar chemistry; Lockyer's extension of the view | 31 |
| Supposed elemental matter in the solar chromosphere | 32 |
| Chemistry of nebulæ; an interstellar medium | 33 |
| Genesis by condensation from primal matter restated | 35 |
| The view reiterated by Mills and by Crookes | 36 |

## CHAPTER IV.

### GASES, LIQUIDS, AND SOLIDS.

| | |
|---|---:|
| Relations between the densities of gaseous and solid species | 38 |
| Polymerism and high equivalents of solids | 39 |
| The condensation in liquid species | 40 |
| Liquids and solids as polymers of vapors | 41 |
| Possibility of fixing their equivalent weights | 42 |

## CHAPTER V.

### THE LAW OF NUMBERS.

| | |
|---|---:|
| Homologous or progressive series in chemistry | 43 |
| Organic chemistry the chemistry of carbon | 44 |
| Extension of the conception of progressive series | 44 |
| Isomeric and anisomeric homologues | 45 |
| Significance of types and structural formulas | 45 |
| The bearings of the periodic law | 46 |

*Contents.*  xi

## CHAPTER VI.

### EQUIVALENT WEIGHTS.

| | |
|---|---|
| The conception of polycarbonates and polysilicates | 47 |
| Their elevated equivalent weights | 47 |
| High equivalents in ammonio-cobalt salts and polytungstates | 48 |
| Complex inorganic acids of Gibbs | 49 |
| Summary of the author's views in 1853 | 49 |

## CHAPTER VII.

### HARDNESS AND CHEMICAL INDIFFERENCE.

| | |
|---|---|
| Density, hardness, and chemical indifference related | 51 |
| Their connection with condensation | 51 |
| The doctrine as stated in 1863 | 52 |
| Relations between chemical and physical characters | 53 |
| Types of mineral silicates; porodic bodies | 54 |
| Hardness and indifference in oxyds and metallic species | 55 |
| Action of fluorhydric acid on silicates | 56 |
| Studies of J. B. Mackintosh thereon | 57 |

## CHAPTER VIII.

### THE ATOMIC HYPOTHESIS.

| | |
|---|---|
| Two hypotheses as to the constitution of matter | 60 |
| The atomic hypothesis down to Dalton's time | 60 |
| The kinetic theory of gases | 61 |
| Whewell on the difficulties of the atomic hypothesis | 61 |
| Its relations to chemical theory | 62 |
| The question of so-called molecular volumes | 63 |
| They are the reciprocals of the coefficient of condensation | 64 |
| Misconceptions as to the chemical process | 65 |
| Extension of the atomic hypothesis from dynamics to chemistry | 66 |
| Its inapplicability to chemical phenomena | 66 |

## CHAPTER IX.

### THE LAW OF VOLUMES.

| | |
|---|---|
| The doctrine of chemical equivalents based on volume | 68 |
| The great generalization of Gay Lussac | 69 |
| All things are by measure, and number, and weight | 69 |
| Universality of the law of volumes | 70 |
| Condensation from the gaseous to the solid state | 70 |
| Supposed relations of volume to isomorphism | 71 |
| Conclusions thereon of Dumas and others | 72 |
| The significance therein of crystalline form questioned | 73 |
| The coefficient of condensation determined | 75 |
| Relations of density between steam and water | 76 |
| Relations between hydrogen, steam, and water | 77 |
| Equivalent weights of water and of ice | 78 |
| An expression sought for the volume | 79 |

## CHAPTER X.

### METAMORPHOSIS IN CHEMISTRY.

| | |
|---|---|
| Chemism defined; metagenesis and metamorphosis | 80 |
| Metamorphosis as polymerization and depolymerization | 81 |
| Fusion, vaporization, and condensation are chemical changes | 81 |
| Arbitrary chemical units for non-volatile species | 82 |
| Metamorphoses of aldehyde | 83 |
| Chloral and methylene oxyd | 84 |
| The various turpentine-oils | 85 |
| Pentine and its polymerization | 86 |
| The heat evolved in such changes | 87 |
| Exemplified by metallic oxyds and silicates | 88 |
| Metamorphosis of sulphur-vapor | 89 |
| Polymerism of vapors under pressure | 90 |
| Various liquid polymers; forms of phosphorus | 91 |
| Metamorphosis of tin | 92 |
| Oxygen and ozone; iodine and chlorine | 93 |

| | |
|---|---|
| Dissociation a universal law | 94 |
| Stages in chemical metamorphosis | 95 |
| Determination of the unit in polymerization | 96 |
| Illustrations of condensation | 97 |
| Species soluble and insoluble in water | 98 |

## CHAPTER XI.

### THE LAW OF DENSITIES.

| | |
|---|---|
| The densities of calcium carbonate examined | 100 |
| The revised combining weight of oxygen | 101 |
| Variations in density of related species | 102 |
| Breithaupt's studies of the calcites | 103 |
| Their calculated formulas and densities | 104 |
| The principle of crystalline admixture | 106 |
| Relations in density of gases and solids | 107 |
| The reciprocal of the coefficient of condensation | 108 |
| Densities of water, steam, and hydrogen compared | 109 |
| The density of the hydrocarbon butane | 110 |
| The formula $p \div d = v$ considered | 110 |

## CHAPTER XII.

### A HISTORICAL RETROSPECT.

| | |
|---|---|
| The doctrine of high equivalents and complex formulas | 112 |
| Conclusions of Favre and Silbermann | 113 |
| Graham on polymerism of dissolved salts | 113 |
| Spencer Pickering on molecular weights | 114 |
| Guthrie on molecular equivalents | 115 |
| Tilden and Shenstone on solution at high temperatures | 116 |
| Louis Henry on the polymerization of oxyds | 118 |
| Density as a function of equivalent weight | 120 |
| Henry's inquiry for the coefficient of polymerization | 120 |
| He restates the author's argument of 1853 | 121 |

Oxyds classified with regard to polymerization . . . . . . . 121
Henry Wurtz's geometrical chemistry . . . . . . . . . . . 122
His hypothesis of varying molecular diameters . . . . . . 124
His mode of calculating densities examined . . . . . . . . 125
Graham on gaseous and liquid diffusion . . . . . . . . . . 128
His scale of solution-densities . . . . . . . . . . . . . . 130
His complex or polymeric molecules . . . . . . . . . . . 130
Non-diffusible bodies, or colloids . . . . . . . . . . . . 132
Favre and Silbermann on some chemical changes . . . . . 133
Henri Sainte-Claire Deville on dissociation . . . . . . . . 133
Dissociation compared to evaporation . . . . . . . . . . . 134
Diffusion in gaseous dissociation . . . . . . . . . . . . 135
Combination compared to condensing of vapors . . . . . . 136
Connections and consequences of the work of Graham and Deville, 137

# CHAPTER XIII.

## CONCLUSIONS.

The chemical process again defined . . . . . . . . . . . . 139
Heterogeneous and homogeneous changes . . . . . . . . 140
Metamorphosis and polymerization . . . . . . . . . . . . 141
Varying stability of polymers . . . . . . . . . . . . . . 142
Metamorphoses of so-called elements . . . . . . . . . . . 143
Genesis of elemental species . . . . . . . . . . . . . . . 143
Changes of state considered . . . . . . . . . . . . . . . 144
Condensation as related to hardness and insolubility . . . . 145
The law of volumes considered . . . . . . . . . . . . . . 146
Relations of water, steam, and hydrogen gas . . . . . . . 147
The proportion $d:p::1:v$ re-examined . . . . . . . . 148
The coefficient of condensation . . . . . . . . . . . . . 149
Density of solids and liquids a function of equivalent weight . 149
The atomic hypothesis in its relations to chemistry . . . . . 150
Pressure as related to chemical change . . . . . . . . . . 151
Heat as the universal disintegrator . . . . . . . . . . . . 151
Relations of chemistry to dynamics and biotics . . . . . . 152

*Contents.*

## CHAPTER XIV.

### SUPPLEMENT.

| | |
|---|---|
| Farther list of the author's papers on chemical theory | 153 |
| The question of homologous or progressive series | 155 |
| Intermediate oxyds, sulphids, and arsenids | 159 |
| The doctrine of valency or atomicity considered | 162 |
| Definite proportions; Cooke, Schützenberger, and Boutlerow | 166 |
| Cooke on the atomic hypothesis in chemistry | 169 |
| Chemical condensation; the dissociation of elemental vapors | 172 |
| Chemical integers and integral weights | 174 |
| The history of chlorhydric and fluorhydric acids | 176 |
| Polymerism and allotropism considered | 177 |
| Cagniard de Latour, Andrews, and Miller on gasefaction | 178 |
| Continuity of the gaseous and liquid states of matter | 182 |
| Solubility of solids in gases; Hannay and Hogarth | 187 |
| The nature of dense vapors discussed; Ramsay and Young | 191 |
| H. Deville on evaporation and dissociation | 195 |
| Correlation of mechanical and chemical forces; Sorby | 197 |
| Effects of great pressure on solids; W. Spring | 199 |
| Heterogeneous integration; metamorphosis; elasticity | 201 |
| Continuity of liquid and solid states of matter | 207 |
| Chemical and dynamical processes distinguished | 209 |
| Heat as an agent favoring integration | 211 |
| Dissociation by electricity and by reduced pressure | 213 |
| Grünewald on hydrogen and solar dissociation | 215 |
| Hydrogen gas as the unit of specific gravity | 216 |
| Coefficients of expansion for water and for non-volatile solids | 221 |
| Calculation of the integral weights of liquids and solids | 223 |
| Significance of the so-called molecular volume | 224 |
| Arbitrary unit-values for $p$ in non-volatile solids | 225 |
| A quantivalent chemical notation for such species | 227 |

### APPENDIX.

| | |
|---|---|
| Hardness and chemical indifference | 231 |
| The integral weight of water | 232 |

# A BASIS FOR CHEMISTRY.

## CHAPTER I.

### INTRODUCTION.

§ 1. At an early period in his chemical studies, the writer was led to adopt certain principles which, as said in 1853, would, it was thought, "be found to enlarge and simplify the plan of chemical science," and might "serve as the basis of a sound theory of chemistry," which he believed to be wanting. His principal publications on this subject, from 1848 to 1886, seventeen in number, are as follows:—

1. ON SOME ANOMALIES IN THE ATOMIC VOLUMES OF SULPHUR AND NITROGEN, WITH REMARKS ON CHEMICAL CLASSIFICATION; American Journal of Science for September, 1848 (vi. 170–178).

2. ON SOME PRINCIPLES TO BE CONSIDERED IN CHEMICAL CLASSIFICATION; read at the first

meeting of the American Association for the Advancement of Science, Philadelphia, September, 1848, and published in the American Journal of Science in 1849 (vol. vii. 399–405; viii. 89–95).

**3.** A short treatise on ORGANIC CHEMISTRY, forming Part IV. (pp. 377–538) of the First Principles of Chemistry by B. Silliman, Jr., third edition, 1852. The novel points in chemical theory in this treatise are resumed by the writer in the American Journal of Science for 1853 (vol. xv. 150–152), and set forth more at length in the next paper mentioned.

**4.** ON THE THEORY OF CHEMICAL CHANGES AND ON EQUIVALENT VOLUMES; published in March, 1853, in the American Journal of Science (xv. 226–234), reprinted in the same year in the London, Edinburgh, and Dublin Philosophical Magazine, and in a German translation in the Chemisches Centralblatt (1853, page 849).

**5.** ON THE CONSTITUTION AND EQUIVALENT VOLUME OF SOME MINERAL SPECIES; in the American Journal of Science for September, 1853 (xvi. 203–218).

**6.** ILLUSTRATIONS OF CHEMICAL HOMOLOGY; read before the American Association for the Advancement of Science, at Washington, May, 1854; published in its Transactions for that year (pages 237-247), and in part in the last named journal for September, 1854 (xviii. 269-271).

**7.** THOUGHTS ON SOLUTION AND THE CHEMICAL PROCESS; in the American Journal of Science for January, 1855 (xix. 100-103), and in the Chemical Gazette for the same year, page 90.

**8.** THE THEORY OF TYPES IN CHEMISTRY; in the American Journal of Science for March, 1861 (xxxi. 256-264).

**9.** SUR LA NATURE DU JADE; in the Compte Rendu of the French Academy of Sciences of June 29, 1863, and in an English translation by the author in the American Journal of Science for the same year (xxxvi. 426-428).

**10.** THE OBJECT AND METHOD OF MINERALOGY; read before the American Academy of Sciences in Boston, January, 1867, and published in the American Journal of Science for May of that year (xliii. 203-206).

**11. The Chemistry of the Primeval Earth;** a lecture before the Royal Institution of Great Britain, London, May 31, 1867, published in the Proceedings of the Institution, and the Chemical News of June 21, 1867, several times reprinted, and translated into French in Les Mondes.

**12. A Century's Progress in Theoretical Chemistry;** being an address delivered at the grave of Priestley, in Northumberland, Pennsylvania, on the Centennial of Chemistry, July 31, 1874, and published in the American Chemist for August and September of that year.

**13. The Chemical and Geological Relations of the Atmosphere;** published in the American Journal of Science for May, 1880 (xix. 349-363).

**14. The Domain of Physiology; or Nature in Thought and Language;** read before the National Academy of Sciences, Washington, April 18, 1881, and published in the London, Edinburgh, and Dublin Philosophical Magazine for October of that year (V. xii. 223-253).

**15. Celestial Chemistry from the Time**

of NEWTON; read before the Philosophical Society of Cambridge, England, Nov. 28, 1881, and published in its Proceedings (vol. IV. part iii); in the Chemical News, and in the American Journal of Science for February, 1882 (xxiii. 123–133).

**16.** A NATURAL SYSTEM IN MINERALOGY, WITH A CLASSIFICATION OF SILICATES; read before the National Academy of Sciences, at Washington, in April, 1885, and before the Royal Society of Canada, at Ottawa, in May, 1885, and published in the Transactions of the latter Society for the same year (vol. III. part iii. pp. 25–93).

**17.** THE LAW OF VOLUMES IN CHEMISTRY; published in Science for September 10, 1886.

Of the above papers, 4, 7, 8, 10, and 11, with portions of 5, 6, and 9, were reprinted in the author's Chemical and Geological Essays (first and second editions), in 1874 and 1878; while 13, 14, 15, and 16 appear, with some additions, in his Mineral Physiology and Physiography, in 1886. A general view of all these was embodied in two papers, entitled: A BASIS FOR CHEMISTRY; and HARDNESS AND CHEMICAL

Indifference in Solids;—read before the National Academy of Sciences, in Boston, November 9 and 11, and noticed in Science for November 19, 1886.

§ 2. The fundamental points in the author's system were already set forth in the earlier papers, prior to 1855; since which time great advances have been made in dynamical and chemical knowledge. The results of all these lead the writer to conclude that the principles laid down by him a generation since should be again brought forward, and presented in a concise form to the attention of the scientific public, with such additions as are required to complete the philosophy of chemistry then first enunciated.

The propositions which were advanced in the various publications named are here set forth, in large part in language quoted from these, in a series of chapters. In making such quotations, reference is made by appended numbers corresponding to those prefixed in the preceding list.

# CHAPTER II.

### NATURE OF THE CHEMICAL PROCESS.

§ 3. IN inquiring into the philosophy of chemistry, it was said in 1853: "We commence by distinguishing between those phenomena which belong to the domain of physics and those which belong to the chemical history of matter. Under the first head we have, besides the gravity of matter in the abstract, its various conditions of consistence, shape, and volume, with the relation of the latter to weight, constituting specific gravity, and the relations of heat, light, electricity, and magnetism." These not only "modify physically the specific characters of matter, but they have besides important relations to those higher processes which give rise to new species by a complete change in the specific phenomena of bodies. In the capacity of such changes consists the chemical activity of matter." Proceeding, then, to discuss the question of the origin of new species, and the relations of individuals, it was said: "That mode of generation which produces individuals like the parent can present no analogy to the phenomena under consideration; *metagenesis* or alternate generation, and *metamorphosis*, are, however, to a certain extent, prefigured in the chemical changes of bodies." These terms were then adopted in chemical language; and it was said that metagenesis, or the production of new species where two or more are concerned, "is effected in two ways: by condensation and union on the one hand, and by expansion and division on the other. In the first case, two or more species unite and merge their specific character in those of a new species; in the second case, the process is reversed, and a body breaks up into two or more new species." Of chemical metamorphosis, which embraces the phenomena of polymerism, it was said: "In metamorphosis by condensation only one species is concerned; and in metamorphosis by expansion the result is homogeneous and without specific difference."

§ 4. "The chemical history of bodies is a record of these changes; it is, in fact, their genealogy.... By union we rise to indefinitely higher species; but in division a limit is reached in the production of species which seem incapable of farther division; and these, being regarded as primary or original species, are called chemical elements." The processes of union and division "continually alternate with each other; and a species produced by the first may yield by division species unlike its parents. From this succession results double decomposition, or equivalent substitution, which always involves a union followed by division; although, under the ordinary conditions, the process cannot be arrested at the intermediate stage. In the production of hydrochloric gas from chlorine and hydrogen, union takes place, followed by immediate expansion without specific difference (metamorphosis)"; while in the case of compounds of chlorine with hydrocarbons, which, under changed conditions, break up into hydrochloric acid and a chlorinized species, it was said, "we observe the intermediate stage."

"A body may divide into two or more new species; yet it is evident that these did not preexist in it, from the fact that a different division may yield other species whose pre-existence is incompatible with the last; nor can the pre-existence of any species but those which we have called primary be admitted as possible. . . . For these reasons it is conceived that the notion of pre-existing elements, or groups of elements, should find no place in the theory of chemistry. Of the relation which subsists between the higher species and those derived from them, we can only assert the possibility, and, under proper conditions, the certainty of producing the one from the other. Ultimate chemical analyses and the formulas deduced from them serve to show what changes are possible in any body, or to what new species it may give rise by its changes." (4.)

§ 5. The question of metamorphosis by condensation, or polymerism, noticed above, had already been discussed at some length in 1848, in considering the vapor-density of sulphur, when it was shown that, if regarded as "consisting of three equivalents combined in one, the density of its vapor is no longer an anomaly, as the sulphur-vapor is condensed to one-third its normal bulk, and its equivalent number is 16 × 3 = 48" [32 × 3 = 96, in the present notation]. It was at the same time proposed to consider ozone as a similar triple group (OOO) corresponding to sulphur-vapor (SSS) and to (SOO). (1.) *Post*, § 56.

In the same year these conclusions were referred to in connection with polymerism in hydrocarbonaceous bodies, and allotropism in elements was explained as polymerism connected with a change in chemical relations. It was said : "Those substances which are considered as elementary may change their equivalents, at the same time undergoing a change in their densities; and, as we obtain from the ordinary equivalent and density an idea of 'the volume of the atom,' we say of those forms having an increased density and a corresponding increase of equivalent (so that the atomic volume remains unchanged) that two or more molecules have united in one. This is illustrated in the case of sulphur," and, as was farther shown, in the compounds of mercury, copper, and iron.

"The so-called elements may then possess

different atomic weights, which enjoy a simple relation to each other, and in these different states exhibit very different characters. When we speak of one of these as containing two or three atoms of another form condensed into one, it is only an expression in accordance with previously existing ideas. We can no longer attach to the atomic weights of the supposed elements an absolute value [that is, as being the weight of an absolutely elemental species]; and thus one of the characteristics which serve to distinguish them from known compounds is rendered of no importance." In farther illustration of this conception, it was then and there suggested that charcoal, graphite, and diamond being "polymeric modifications of elemental carbon," the first of these "is a species of anhydrid derived from cellulose"; and, moreover, that the allotropic red phosphorus might be a similar polymer;—a question which, it was said, "would be solved by a determination of its density in that state." (2.)

§ 6. In 1854, returning to the subject of chemical change, and discussing the question of double decomposition, it was illustrated by the production of arsenious oxyd and chlorhydric acid by the action of water on terchlorid of arsenic, when it was said: "This decomposition of the solution of chlorid of arsenic is an example of what is generally called double elective affinity (*attractio electiva duplex*), and is generally explained by saying that the attraction of arsenic for oxygen, and that of chlorine for hydrogen, enable the chlorid and the water to decompose each other. But these elemental species do not exist in the solution, although they are possible results of its decomposition; and to explain the process in this manner is to ascribe it to the affinities of yet unformed species."

It was farther said that "double decomposition always involves union followed by division; although we cannot in every case arrest the process at its first stage. Under some changed conditions of temperature and pressure the decomposition may be counterpart of the previous union"; as in the example then given of mercury and oxygen. "When the division takes place in a sense different from the union, giving rise to new species, we have double decomposition. .... It is only when looked upon as a momentary combination, followed by a decomposition, that the theory of double decomposition becomes intelligible and is in accordance with known facts. From the narrow limits of temperature which often include the processes, and from the ease with which light, warmth, friction, and pressure excite the decomposition of such bodies as the chlorid of nitrogen, the nitrite of ammonia, the oxyds of chlorine, and the metallic fulminates, we may conclude that within still narrower limits, and under conditions as yet undefined, many bodies may exhibit affinities for each other which are reversed by a very slight change of condition. In this way we may explain many of those obscure phenomena hitherto ascribed to *action by presence* or *catalysis.*" (**7**.)

§ 7. The above doctrines were restated in 1861, as follows: "All chemical changes are reducible to union (identification) and division (differentiation). When in these changes only one species is concerned, we designate the process as metamorphosis, which is either by condensation or by expansion (homogeneous differentiation). In metagenesis, on the contrary, unlike species may unite and, by a subsequent heterogeneous differentiation, give rise to new species, constituting what is called double decomposition; the results of which, differently interpreted, have given rise to the hypothesis of radicles, and the notion of substitution by residues, to express the relations between the parent bodies and their progeny. The chemical history of bodies is a record of these changes; it is, in fact, their genealogy, and in making use of typical formulas to indicate the derivation of chemical species, we should endeavor to show the ordinary modes of generation." (8.)

§ 8. In farther exposition of the nature of the chemical process, it was said in 1853: "Chemical species are homogeneous; *tota in minimis existit natura.*" "Chemical combination is not a putting together of molecules, but an interpenetration of masses." "Chemical union is interpenetration, as taught by Kant, and not juxtaposition, as conceived by the atomistic chemists. When bodies unite, their bulks, like their specific characters, are lost in those of the new species." (4.) In the year 1855 it was farther said of Kant's definition, that "the conception is mechanical, and therefore fails to give an adequate idea. The definition of Hegel, that *the chemical process is an identification of the different and a differentiation of the identical*, is, however, completely adequate. Chemical union involves an identification not only of the volumes (interpenetration mechanically considered), but of the specific character of the combining bodies, which are lost in those of the new species." (7.) In farther illustration of the above conception, in 1867, in referring to the speculations of Macvicar and Gaudin, as to "the architecture of crystalline molecules," it was written: "Nature builds up her units by interpenetration and identification, and not by juxtaposition of the chemical elements." (10.)

§ 9. In 1874, in again discussing this question, it was said: "In chemical change the uniting bodies come to occupy the same space at the same time, and the impenetrability of matter is seen to be no longer a fact. The volume of the combining substances is confounded, and all the physical and physiological characters which are our guides in the region of physics fail us, gravity alone excepted. The diamond dissolves in oxygen, and the identities of chlorine and sodium are lost in that of sea-salt. To say that chemical union is, in its essence, identification, as Hegel has defined it, seems to me the simplest statement conceivable. The type of the chemical process is found in solution, from which it is possible, under changed physical conditions, to regenerate the original species." (12.)

§ 10. Chemical change, then, may be defined as an integration or a disintegration of chemical species, resulting in the genesis of new species, which are themselves chemical; that is to say, mineral, and not organic. All of these "may be supposed to be formed from a single element, or *materia prima*, by the chemical process." "The chemical species, until it attains to individuality in the crystal, is essentially quantitative." (10.) Gases, liquids, and colloids have a specific existence, but no individuality; mechanical subdivision does not destroy them. "The activities of the crystal are purely dynamic, and its crystalline individuality must be destroyed before it can be the subject of chemism, while the plant and the animal exhibit not only dynamical and chemical, but organogenic activities, which last are designated as vital phenomena." All of these are necessary for the preservation of the organism. The study of these vital activities constitutes a third division of physics, which we have elsewhere called biotics.

"Mineralogy is the science of inorganic matter, and studies its dynamical and chemical relations; while biology, which is the science of organic matter, adds to these the study of biotical relations. The dynamical and chemical activities which, in the mineral kingdom, give rise to the crystalline individual, are therein in static equilibrium. The organic individual, on the contrary, is kinetic, and maintains its equilibrium by perpetual adjustment with the outer world." "The physiology of matter in the abstract is dynamical; that of mineral forms is both dynamical and chemical; while that of organic forms is at once dynamical, chemical, and biotical." (14.)[1]

§ 11. By keeping in mind the above definitions, we shall avoid the error of modern students, who are disposed to confound dynamic activities with chemism itself, and thus to lose sight of the essential nature of the chemical process, which, as was pointed out in 1853 (§ 3), is to be clearly distinguished from

[1] These relations are shown by the annexed tabular view of the classification of the natural sciences, revised from the writer's Mineral Physiology and Physiography, page 29: —

| | | | | INORGANIC NATURE. | ORGANIC NATURE. |
|---|---|---|---|---|---|
| NATURAL SCIENCES. | DESCRIPTIVE. | *General Physiography* or | *Natural History.* | MINERAL PHYSIOGRAPHY. ——— Astronomy, descriptive. Mineralogy, descriptive and systematic. Geognosy. Geography. | BIOPHYSIOGRAPHY. ——— Organography. Botany and Zoölogy, descriptive and systematic. |
| | PHILOSOPHICAL. | *General Physiology* or | *Natural Philosophy.* | MINERAL PHYSIOLOGY. ——— *Dynamics. Chemistry.* Astronomy, theoretical. Mineralogy, physiological. Geogeny. | BIOPHYSIOLOGY. ——— *Biotics.* Organogeny. Morphology. Botany and Zoölogy, physiological. |

the phenomena belonging to what we may call dynamism. As examples of this prevailing confusion, it was said, in 1881: "Clifford wrote of molecular motion, 'which makes itself known as light, or radiant energy, or chemical action'; while Faraday was wont 'to express his conviction that the forces termed chemical affinity and electricity are one and the same.' Helmholtz, from whom I here quote, adds: 'I think the facts leave no doubt that the very mightiest among the chemical forces are of electrical origin . . . but I do not suppose that other molecular forces are excluded, working directly from atom to atom.'" Similar language might be quoted from other not less eminent authorities, to show the vague notions still reigning in the minds of modern students as to the scope of chemistry, and its relation to dynamics.

"The activities which appear in dynamical and in chemical phenomena are one in essence, for force is one. The same is true of the activities manifested in organic growth, and even in thought; but the unity and mutual convertibility of different manifestations of force afford no ground for confounding, as some would do, dynamics with chemism, or with vital or mental processes. All these phenomena are but evidences of universal animation; or, in other words, of an energy which is inherent in matter, the manifestations of which, as matter rises to higher stages of development, become more complex, as organic individuals are themselves more complex than mineral forms."

"When the energy which is in matter is manifested without reference to species, we call it simply dynamics; when it results in the production of mineral species, we call it chemics, or chemism; and when it gives rise to organisms, which may be defined as kinetic individuals, we distinguish it as vital, or biotic. In matter we must recognize, with Tyndall, 'the promise and the potency of all terrestrial life.'" (14.) Those dynamic activities which manifest themselves as electricity, temperature, and radiant energy, while attendant upon chemical processes, are yet to be as carefully distinguished from the essential phenomena of chemism as the latter are from those of biotics.

§ 12. In farther illustration of the nature of chemical combination, it was said, in 1853:

"Solution is a result of that tendency in nature which constantly leads to unity, condensation, identification." "Solution is chemical union, as is indicated by the attendant condensation." (**4.**) Again, in 1855, it was written: "All chemical union is nothing else than solution; the uniting species are, as it were, dissolved in each other; for solution is mutual." "Solution, then, being identification, the discussion as to whether metallic chlorids are changed into hydrochlorates when dissolved in water is meaningless. Such a solution is a unity, in which we can no more assert the existence of the chlorid, or of water, than of chlorine, hydrochloric acid, or a metallic oxyd; although these, and many others, are conceivable results of its differentiation." (**7.**) In the words already cited, —"The type of the chemical process is found in solution, from which it is possible, under changed physical conditions, to regenerate the original species." (**12.**)

# CHAPTER III.

## GENESIS OF THE CHEMICAL ELEMENTS.

§ 13. THERE remains to be considered another aspect of the chemical process — namely, the production, from a primal undifferentiated matter, of the chemical elements, as suggested by Lorenz Oken.[1] "The successive steps in

---

[1] In this connection we quote from page 3 of Tulk's Translation of Oken's Physiophilosophy, published by the Ray Society in 1847, with a preface by Oken himself, the following passage: "Physiophilosophy is divisible, therefore, into three parts. . . . The first division is the doctrine of the Whole (*de Toto*) — Mathesis. The second that of Singulars (*de Entibus*) — Ontology. The third that of the Whole in Singulars (*de Toto in Entibus*) — Biology. The science of the Whole must divide itself into two doctrines; into that of immaterial totalities — Pneumatogeny; and that of material totalities — Hylogeny. Ontology teaches us the phenomena of matter. The first phenomena of this are the heavenly bodies, comprehended by Cosmogony. These develop themselves further, and divide [differentiate] into the elements — Stoichiogeny. From these the earth-element develops itself still further, and divides into minerals — Mineralogy. These minerals unite in one collective body, and this is Geogeny. The Whole in Singulars is the living or Organic, which again divides into plants and animals." The study of these includes Biology, with its subdivisions.

the ontological process are, first, Cosmogony, or the fashioning of the heavenly bodies from the previously formed matter; followed by the genesis therefrom of the chemical elements, Stoichiogeny." (14.)

§ 14. The conception of the genesis of the so-called chemical elements first took a definite form from the study of the equivalent weights of related elements, and from the law of numbers made apparent in hydrocarbonaceous bodies. Of this we wrote as follows, in 1874, in the essay on A Century's Progress in Theoretical Chemistry (12) : —

"As early as 1829, Dobereiner called attention to the triads of related bodies, such as the groups of lithium, sodium, and potassium, and of calcium, strontium, and barium, in which the equivalent weight of the middle term is the mean of that of the extremes. Almost simultaneously, and apparently independently, Pettenkofer and Dumas took up again the question, in 1851, and noticed the fact that the numerical relations between these and many other mineral radicles are similar to those between the organic radicles known to result from the condensation of successive equivalents of carbon and hydrogen, such as the olefines, the paraffines, and the alcohol radicles. The possibility that the mineral radicles, related alike in chemical characters and in equivalent weights, might, also, be composite bodies, was suggested by Dumas in 1857; and with this in view he undertook a re-examination of the equivalent weights of the elements. If, as the numbers adopted by Berzelius seemed to show, no simple ratios existed between those of related elements, the suggestion of Dumas was inadmissible; but if, on the contrary, as supposed by Prout, a simple and exact relation of this kind was established, it was not impossible that the view of their compound nature might be true."

"The results of many years of patient labor devoted to a re-examination of the equivalent weights of a great number of bodies, were given to the world by Dumas, in 1859; and from these he concluded that the law of Prout is so far true that the equivalent weights of the elements are multiples of a unity which is one-fourth that of hydrogen. . . . These results, and the analogies between the elements of mineral and of organic chemistry, both in chemical relations and in equivalent weights (which in the mean time had been carefully investigated and developed by J. P. Cooke), now led Dumas to reaffirm his conjecture that the mineral radicles may really be compounds, though their decomposition is perhaps beyond the possibilities of our chemical analysis." (12.)

§ 15. The next step in the development of the doctrine of Stoichiogeny, so far as known to the writer, is to be found in his lecture of 1867 (11), quoted below. The history thereof, and of certain views enunciated almost simultaneously by Brodie of Oxford and the present writer, and subsequently developed and extended by the latter, is given at length in his essay on Celestial Chemistry from the Time of Newton (15), as follows:—

"In Part I. of his Calculus of Chemical Operations, read before the Royal Society, May 3, 1866, and published in the Philosophical Transactions for that year, Brodie was led to assume the existence of certain ideal elements. These, he said, 'though now revealed to us through the numerical properties of chemical equations only as *implicit and dependent existences*, we cannot but surmise may sometimes become, or may in the past have been, *isolated and independent existences.*' Shortly after this publication, in the spring of 1867, I spent several days in Paris with the late Henri Sainte-Claire Deville, repeating with him some of his remarkable experiments in chemical dissociation, the theory of which we then discussed in its relations to Faye's solar hypothesis. From Paris, in the month of May, I went, as the guest of Brodie, for a few days to Oxford, where I read for the first time, and discussed with him, his essay on the Calculus of Chemical Operations, in which connection occurred the very natural suggestion that his ideal elements might, perhaps, be liberated in solar fires, and thus be made evident to the spectroscope."

§ 15A. "I was then about to give, by invitation, a lecture before the Royal Institution, on The Chemistry of the Primeval Earth, which was delivered May 31, 1867. A stenographic report of the lecture, revised by the author, was published in the Chemical News of June

21, 1867, and in the Proceedings of the Royal Institution. Therein, I considered the chemistry of nebulæ, sun, and stars in the combined light of spectroscopic analysis and Deville's researches on dissociation, and concluded with the generalization that the 'breaking-up of compounds, or dissociation of elements, by intense heat is a principle of universal application, so that we may suppose that all the elements which make up the sun, or our planet, would, when so intensely heated as to be in the gaseous condition which all matter is capable of assuming, remain uncombined; that is to say, would exist together in the state of chemical elements; whose further dissociation in stellar or nebulous masses may even give us evidence of matter still more elemental than that revealed in the experiments of the laboratory, where we can only conjecture the compound nature of many of the so-called elementary substances' (11)."

§ 16. "The importance of this conception, in view of subsequent discoveries in spectroscopy and in stellar chemistry, has been well set forth by Lockyer, in his late lectures on Solar Physics,[1] where, however, the generalization is described as having been first made by Brodie, in 1867. A similar but later enunciation of the same idea by Clerk-Maxwell is also cited by Lockyer. Brodie, in fact, on the 6th of June, one week after my own lecture, gave a lecture on Ideal Chemistry before the Chemical Society of London, published in the Chemical News of June 14, in which, with regard to his ideal elements (in further extension of the suggestion already put forth by him in the extract above given from his paper of May 6, 1866), he says: 'We may conceive that in remote ages the temperature of matter was much higher than it is now, and that these other things [the ideal elements] existed in the state of perfect gases — separate existences — uncombined.' He further suggested, from spectroscopic evidence, that it is probable that 'we may one day, from this source, have revealed to us independent evidence of the existence of these ideal elements in the sun and stars.' During the months of June and July, 1867, I was absent on the Continent;

[1] Nature, Aug. 25, 1881, vol. xxiv. p. 396.

and this lecture of Brodie's remained wholly unknown to me until its republication in 1880, in a separate form, by its author,[1] with a preface, in which he pointed out that he had therein suggested the probable liberation of his ideal elements in the sun, referring at the same time to his paper of 1866, from which we have already quoted the only expression bearing on the possible independence of these ideal elements somewhere in time or in space."

"The above statements are necessary in order to explain why it is that I have made no reference to Sir Benjamin Brodie on the several occasions on which, in the interval between 1867 and the present time [1881], I have reiterated and enforced my views on the great significance of the hypothesis of celestial dissociation as giving rise to forms of matter more elemental than any known to us in terrestrial chemistry. The conception, as at first enunciated, in somewhat different forms, alike by Brodie and myself, was one to which we were both naturally, one might say inevitably, led, by different paths, from our respective fields of

[1] Ideal Chemistry, a Lecture. Macmillan, 1880.

speculation, and which each might accept as in the highest degree probable, and make, as it were, his own. I write, therefore, in no spirit of invidious rivalry with my honored and lamented friend, but simply to clear myself from the charge, which might otherwise be brought against me, of having, on various occasions within the past fourteen years, put forth and enlarged upon this conception without mentioning Sir Benjamin Brodie, whose only publication on the subject, so far as I am aware, was his lecture of 1867, unknown to me until its reprint in 1880."

§ 17. "It was at the grave of Priestley, in 1874, that I for the second time considered the doctrine of celestial dissociation, commencing with an account of the hypothesis put forward by F. W. Clarke, of Cincinnati, in January, 1873,[1] to explain the growing complexity which is observed when we compare the spectra of the white, yellow, and red stars ; in which he saw evidence of a progressive evolution of chemical species, by a stoichiogenic process, from more

[1] Clarke, "Evolution and the Spectroscope," Popular Science Monthly, New York, vol. ii. p. 32.

elemental forms of matter. I then referred to the further development of this view by Lockyer, in his communication to the French Academy of Sciences in November of the same year, wherein he connected the successive appearance in celestial bodies of chemical species of higher and higher vapor-densities with the speculations of Dumas and Pettenkofer as to the composite nature of the chemical elements.[1] I next quoted from my lecture of 1867 the language already cited (on page 28), to the effect that dissociation by intense heat in stellar worlds might give us more elemental forms of matter than any known on earth. It was further suggested that the green line in the spectrum of the solar corona (which had been supposed to indicate a hitherto unknown substance, that, 'from its extension beyond even the layer of partially cooled hydrogen, must, according to the deductions of Mr. George J. Stoney, be still lighter than this gas') may be due to a 'more elemental form of matter, which, though not seen in the nebulæ, is liberated by the intense heat of the solar sphere, and

---

[1] Lockyer, Comptes Rendus, Nov. 3, 1873.

## Genesis of the Chemical Elements. 33

may possibly correspond to the primary matter conjectured by Dumas,[1] having an equivalent weight one-fourth that of hydrogen.' (12.) The suggestion of Lavoisier, that 'hydrogen, nitrogen, and oxygen, with heat and light, might be regarded as simpler forms of matter from which all others are derived,' was also noticed in connection with the fact that the nebulæ, which we conceive to be condensing into suns and planets, have hitherto shown evidences only of the presence of the first two of these elements; which, as is well known, make up a large part of the gaseous envelope of our planet, in the forms of air and aqueous vapor. With this I connected the hypothesis that our atmosphere and ocean are but portions of the universal medium which, in an attenuated form, fills the interstellar spaces; and further suggested as 'a legitimate and plausible speculation,' that 'these same nebulæ and their resulting worlds may be

[1] In the paper of 1874, it was farther said in this connection: "Mention should also be made of the unknown element conjectured by Huggins to exist in some nebulæ. This conception of a first matter, or *Urstoff*, has also been maintained by Hinrichs, who has put forward an argument in its favor from a consideration of the wave-lengths in the lines of the spectra of various elements."

evolved by a process of chemical condensation from this universal atmosphere, to which they would sustain a relation somewhat analogous to that of clouds and rain to the aqueous vapor around us.'"

§ 18. "These views were reiterated in the preface to a second edition of my Chemical and Geological Essays, in 1878, and again before the British Association for the Advancement of Science, at Dublin,[1] and before the French Academy of Sciences in the same year.[2] They were still further developed in the essay on The Chemical and Geological Relations of the Atmosphere, published in 1880 (13), in which attention was called to the important contribution to the subject by Mr. Lockyer, in his ingenious and beautiful spectroscopic studies, the results of which are embodied in his 'Discussion of the Working Hypothesis that the so-called Elements are Compound Bodies,' communicated to the Royal Society, Dec. 12, 1878. It was then remarked that the already noticed 'speculation

---

[1] Nature, Aug. 29, 1878, vol. xviii. p. 475.
[2] Comptes Rendus, Sept. 23, 1878, vol. xxxviii. p. 452.

of Lavoisier is really an anticipation of that view to which spectroscopic study has led the chemists of to-day'; while it was said that the hypothesis put forth by the writer in 1874, 'which seeks for a source of the nebulous matter itself, is, perhaps, a legitimate extension of the nebular hypothesis.'" (15.)

§ 19. That the law of chemical dissociation by heat is universal, and that in nebulous and stellar masses we may expect to find more elemental forms of matter than are known on earth, was maintained by the writer in May, 1867. In August, 1874, connecting this conception with the earlier speculations of Dumas as to the composite nature of the so-called chemical elements, and with the order of appearance of these in different classes of stars, as noticed by Clarke, it was suggested that the material from which the known elements have been generated by the stoichiogenic process, through successive condensations in cooling, is no other than the unknown element discerned by the spectroscope in the solar chromosphere, as the well known line, 1474. These views, which, as we have shown, have since been persistently and repeatedly asserted by the writer, are now finding recognition among chemists, some of whom seem to overlook the history of their origin. Such a genesis of the elements is assumed by E. J. Mills, who, in 1886, writes:[1] "The cooling of the primitive matter may be regarded from a chemical point of view as resulting in a succession of polymers $(1, 2, 3 \ldots )n;$ $n$ being the primitive density. But, on account of the evolution of heat when a polymer is formed, there will ensue, as a physical consequence, the inversion of more or less of the cooling, and therefore of the polymerization"; a process of which it is easy to trace the history, as Mills has done, in variable stars.

§ 20. In his address before the British Association for the Advancement of Science, in September, 1886, Crookes has given a popular statement of the hypothesis of the genesis of the chemical elements from an intensely heated primal matter, during the cooling of which they have been formed by successive polymeriza-

---

[1] Numerics of the Elements, Part II. L., E., and D. Philos. Mag. xxi. 157.

tions. All of this he connects with Prout's hypothesis of atomic weights, and suggests that an unknown element in the solar atmosphere may be, if not the primal matter, a substance having one-half the weight of hydrogen. I had already in 1874 expressed the opinion that this substance might be the element giving the green line, 1474, of Kirchhoff, so conspicuous in the corona; of which Young[1] remarks that it would seem that "it must be something with a density far below that of hydrogen." Crookes, however, now suggests, in 1886, another element, the hypothetical helium of Frankland, which gives the yellow line $D_3$. Still other unknown lines have of late years been discovered in the solar spectrum, all of which may correspond to so-called elemental species developed by the stoichiogenic process; and it may be a question to which the primacy should be conceded.

[1] C. A. Young, The Sun, 1881; p. 232.

# CHAPTER IV.

### GASES, LIQUIDS, AND SOLIDS.

§ 20. AFTER noting, in 1853, that the equivalents for volatile bodies are fixed from their vapor-densities, and for "non-volatile species are generally assumed to be those quantities which sustain the simplest ratio to certain volatile ones," it was said that, "having determined the true equivalent of a species from the density of its vapor, the inquiry arises whether a definite and constant relation may not be discovered between its vapor-density and the specific gravity of a species in its solid state. Such a relation being established, and the value of the condensation in passing from a gaseous to a solid state being known, the equivalents of solids, like those of vapors, might be determined from their specific gravities." In answer to the question then asked, "What is the value of the condensation which takes place in the change from the gaseous to the solid state, or what equivalent corresponds to a given specific gravity in any crystalline solid?" the received formulas for alum, ferrocyanid of potassium, and glucose were discussed; and it was said: " The equivalent of a crystallized species may often be a multiple of that deduced from those chemical changes which commence only with the destruction of its crystalline individuality." It was added: "There are reasons for believing that the equivalents of these species in the crystalline state correspond to some multiples of the above formulas, a question which is to be decided by an examination of the crystallization and the specific gravity of species whose equivalents are admitted to be higher."

Farther, it was said: "Favre and Silbermann from their researches on the heat evolved in fusion and solution, have been led to conclude, firstly, that crystallized salts are polymeric of these same salts in solution — that is, they are represented by formulas which are multiples of those deduced from analysis; secondly, that double salts and acid salts do not exist in solution, being produced only during crystallization; and, thirdly, that water, in crystallizing, changes from HO to $n$HO, $n$ being some whole number.[1] These conclusions are seen to be in accordance with those deduced from a consideration of the relations of density and equivalent volume. A polymerism is evident in such salts as sulphate of potash and cyanid of potassium, when their specific gravities are compared with those of alum and the ferrocyanid."

§ 21. Passing thence to the consideration of the alcohols and their derivatives, it was remarked that several of those then known "have very nearly the same specific gravity, so that the condensation is inversely as their vapor-equivalents." It was added that, in a farther study of such bodies, "the specific gravity at their boiling-points should probably be chosen for comparison," and that we may expect to "establish a simple relation between the densities of liquids and their vapors." (**4**.)

§ 22. In resuming this subject in 1867, it was farther said: "There probably exists be-

---

[1] Recherches sur les quantités de chaleur dégagées dans les actions chimiques et moléculaires. 1847. Comptes Rendus de l'Acad. des Sciences, xxiv. 1081–1090.

tween the true equivalent weights of non-gaseous species and their densities a relation as simple as that between the equivalent weight of gaseous species and their specific gravities." That it was possible to discover such a relation, had already been maintained in 1853, when, as we have seen, the doctrines of polymerism and of high equivalent weights for liquid and solid species had been fully set forth. The chemical relation between solids, liquids, and gases was now, however, first clearly defined as follows :

"The gas or vapor of a volatile body constitutes a species distinct from the same body in its liquid or solid state, the chemical formula of the latter being some multiple of the first; and the liquid and solid species often [probably always] constitute two distinct species, of different equivalent weights. In the case of analogous volatile species, as the hydrocarbons and their derivatives, the equivalent weights of the liquid and solid species approximate to a constant quantity, so that the densities of those species in the case of homologous or related alcohols, acids, ethers, and glycerids, are subject to no great variations. These non-gaseous species are generated by the chemical union or identification of a number of volumes or equivalents of the gaseous species, which number varies inversely as the density of those gaseous species." **(10.)**

In 1874, reverting to the conception of the polymerism exhibited in different forms of carbon and phosphorus, as maintained in 1848 (page 12), it was said: "By the comparison of these with substances known to possess a high equivalent weight, as, for example, some organic bodies, it would even be possible to fix that of these elemental species; which would certainly be found to have a very elevated equivalent, indicating a high degree of polymerism." **(12.)**

# CHAPTER V.

### THE LAW OF NUMBERS.

§ 23. THE law of numbers, which is seen in the doctrine of multiple proportions and in polymerism, received a great extension in the discovery of what have been called progressive series, made known in 1842, by James Schiel, and adopted by Ch. Gerhardt, in 1844, in his Précis de Chimie Organique, under the name of homologous series. This conception was by these chemists applied only to hydrocarbonaceous compounds, differing from each other by $C_2H_2$, which was shown to be the common difference in the formulas of certain series of bodies having similar chemical relations.

As regards the farther extension of this principle, and the breaking-down of the distinction hitherto maintained between so-called organic chemistry and mineral chemistry, it was said by the writer, in 1852, that "we may define organic chemistry as the chemistry of the compounds of carbon." These were then spoken of as "the carbon series"; while "the silicon series" was made to include all the known silicious compounds. (**3.**) Of the above definition it has been elsewhere said that, "though a commonplace to-day, it was, perhaps, then made for the first time." (**16.**)

(**3.**) In 1853 it was farther pointed out that "$C_2H_2$ may be compared with $O_2H_2$ and with $O_2M_2$" ($CH_2$, $OH_2$, $OM_2$, in the present notation); so that since species differing by $n(C_2H_2)$ may be homologous, "it may be expected that mineral species will exhibit the same relations as those of the carbon series, and the principle of homology be greatly extended in its application," as was then shown by reference to many silicates. (**4.**) In the same year it was also said: "The formulas of homologous bodies may be represented as series in arithmetical progression [progressive series]. The first term may be the same as the common difference," as in the hydrocarbons of the series $n(CH_2)$, or unlike the common difference, as in the ammonias $NH_3$. . . . $NH_3.n(CH_2)$. "Both of these cases are illustrated in the chemical history of mineral species"; and "$M_2S_2$, $M_2O_2$, and $H_2O_2$ may be compared with $H_2C_2$." The application of these principles was then shown at length. (5.) The members of a progressive series in which the first term is the same as the common difference may be designated as isomeric homologues; while all other progressive series are appropriately called anisomeric.

§ 24. The progressive series among gaseous species, except in some few hydrocarbons, are examples of anisomeric homologues. To these gaseous species belong all considerations as to metagenesis, from the study of which have grown up the so-called rational and structural formulas, and the theory of types. With regard to all these it is well to remember "the language of Ch. Gerhardt, that they 'are not intended to represent the arrangement of atoms, but to make evident, in the simplest and most direct manner, the relations which connect bodies with one another in their transformations.' This understood, and with these reservations, they have an important place in chemical teaching." (**12.**)

The new aspects revealed by the periodic law, and the discovery of the reciprocal relations between the properties and the combining weights of the elements suffice to show that the hypothesis of Prout must, at least, be greatly modified before it can be received, and help, moreover, to give a more profound significance than was before imagined to the law of numbers in chemistry.

In 1874 the writer noticed the variations in composition which J. P. Cooke, who had observed them in certain alloys, described under the name of allomerism. These alloys were then regarded as "examples of a progressive series of isomorphous compounds of antimony and zinc, of high equivalent, differing from each other by $n\mathrm{Zn}_2$."[1] The bearing of this conception on the recent views of Boutlerow and Schützenberger as to the apparent variability of the law of definite proportions, and on Cooke's discussion thereof,[2] is evident.

[1] Hunt, Chemical and Geological Essays, p. 447.
[2] American Journal of Science (1883), xxvi. 63, 310.

## CHAPTER VI.

### EQUIVALENT WEIGHTS.

§ 25. THE conception of high equivalent weights for liquid and solid species, already indicated in the extracts in §§ 20–22, was illustrated at length in 1853 and 1854, when the carbon-spars were represented in formulas as polycarbonates with from thirty to forty portions of carbon ($C = 6$); and the pyroxenes, amphiboles, and feldspars as polysilicates with from thirty-two to sixty portions of silicon ($Si = 7$). The equivalent weights thus provisionally adopted for these species were confessedly *minima;* and their relations to the higher numbers of which they were supposed to be fractions were left undetermined. (**5, 6.**) Of these elevated equivalent weights and complex formulas it was subsequently declared that they do not show "a deviation from the law of definite proportions," but are "only an expression of that law in a higher form." (**10.**)

§ 26. Reverting, in this connection, to what has been said in § 22, we repeat, that all liquid and solid species, so far as known, may be represented as polymers of some primary species or chemical unit. The minimum equivalent weights of these chemical units, which, by their polymerism, give rise to higher species, must often of necessity be considerable, as in the case of many of the aniline derivatives, the ammonio-cobalt compounds, and the polytungstates. Wolcott Gibbs, who shows that these latter constitute progressive series, finds the common difference, which he terms the "homologizing term," in the phosphotungstates is not less than $2(WO_3 = 464)$, and shows for these heavy polytungstates equivalent weights, not only of 5002, but, in one case, of 20,058; while for the less heavy ammonio-cobalt salts minimum equivalent weights of from 500 to 2500 are deduced. Even these elevated numbers, as we shall endeavor to show, are but fractions of the true equivalents of the crystalline polymeric solids, which, for the species under consideration, vary from about 50,000 to more than 220,000.

§ 27. The proofs which careful analytical studies have afforded of the necessarily high equivalent weights and the complex formulas of the polytungstates, polymolybdates, polyvanadates, and polyphosphates, led Wolcott Gibbs, in 1877, to designate them as salts of "complex inorganic acids," which, according to him, "form a new department of inorganic chemistry." It will be remembered that I had already, in 1853, proclaimed that the whole chemistry of solids and liquids is only intelligible when regarded as a history of just such complex inorganic acids and salts; that the distinction between organic and inorganic chemistry is no longer tenable; that the same principles of homology and polymerism are applicable alike to the bodies of the carbon series and the silicon series; that the native crystalline carbonates, or carbon-spars, are polycarbonates, with equivalent weights of not less than from 1500 to 2500; that the pyroxenes, feldspars, and tourmalines are polysilicates of equally complex constitution, and are represented by formulas which show the

[1] American Journal of Science, 1877; xiv., 61.

existence among them both of polymers, probably homologous, and of anisomeric homologues. These conceptions, all of which were explicitly set forth and defended in 1852 and 1853, underlie the writer's philosophy of the mineral kingdom, as then enunciated, and as persistently maintained to the present date. The inquiry as to the value of the unit in these polymers, and of the common difference between the successive members of these homologous series of polycarbonates and polysilicates — whether it be the simplest admissible chemical unit, or some multiple thereof — is one which will be considered in Chapter XI.

## CHAPTER VII.

### HARDNESS AND CHEMICAL INDIFFERENCE.

§ 28. THAT the specific gravity of solids, like that of gases and vapors, varies as their equivalent weights, will follow from the principle, already laid down in 1853, that chemical union is essentially an identification of volumes; or, in other words, a condensation of many volumes into one. The farther and very rational deduction from this principle, that the property of hardness in similarly constituted species, as well as their resistance to chemical agents, sustains a direct relation to the condensation, was first indicated in a paper by the writer, on Euphotide and Saussurite, in the American Journal of Science for 1859, wherein were discussed the physical and chemical differences of the related silicates, meionite and zoisite. In referring to this in a note, "Sur la Nature du Jade," presented to the French Academy of

Sciences in June, 1863, it was said, in comparing meionite and zoisite: "The augmentation of density, of hardness, and of chemical indifference which is seen in this last species is doubtless to be ascribed to a more elevated equivalent; or, in other words, to a more condensed molecule.[1] These different degrees of condensation, which are constantly kept in mind in the study of organic chemistry, are besides, as I have already elsewhere shown, of great importance in mineralogy, and will form the basis of a new system of classification, which will be at once chemical and natural-historical. The different rhombohedral carbon-spars, cyanite and sillimanite, hornblende and pyroxene, offer, in like manner, examples of different degrees of condensation, and, by their chemical composition, belong to series the terms of which, like those of the hydrocarbons — $n(C_2H_2)$ — are both homologues and

[1] Regarding the relations between meionite and zoisite it was said: "L'augmentation de densité, de dureté, et d'indifférence chimique qu'on remarque dans cette dernière espèce tient sans doute à un équivalent plus elévé, c'est à dire à une molécule plus condensée." Comptes Rendus de l'Academie des Sciences, 1863, lvi. 1256.

multiples of the first term. At the same time, each one of these silicates and carbonates belongs to another possible series, the terms of which differ by $n(M_2O_2)$, corresponding to more or less basic salts." (**9**.)

§ 29. These relations of physical and chemical characters to condensation were again affirmed in 1867, when, after insisting on the evidences of polymerism "in related mineral species, such as meionite and zoisite, dipyre and jadeite, hornblende and pyroxene, calcite and aragonite, opal and quartz, in the zircons of different densities, and in the various forms of titanic oxyd and of carbon," it was said: "The hardness of these isomeric or allotropic species, and their indifference to chemical reagents increase with their condensation; or, in other words, vary inversely as their empirical equivalent volumes, so that we here find a direct relation between chemical and physical properties." (**10.**)

§ 30. The same relation of hardness to condensation is evident throughout natural silicates, and underlies the great distinction made in the mineralogical classification of

Mohs between the order Gem, on the one hand, and the order Spar, on the other. This distinction is made still more apparent in the division which I have made the ground of tribal subdivisions running through the three sub-orders of silicates, and in each separating them into the gem-like or adamantoid type, on the one hand, and the spathoid and hydrospathoid types, on the other. Passing from these three, which are all alike essentially sparry in crystalline structure, to the phylloid or micaceous type, the significance of the varying condensation in species of this type, as regards hardness, is somewhat obscured by the eminent cleavage in one plane. The same is true, for another reason, in the colloid, vitreous, or porodic type[1] (*Porodini* of Breithaupt), the species

---

[1] The term porodic (German *porodisch*), from the Greek πωρόω, to harden, coagulate, or make callous, was proposed by Breithaupt, in 1836 (Handbuch der Mineralogie, I. 324), as synonymous with the German *geronnen* (curdled or clotted), to designate amorphous, opal-like, gelatinous, or vitreous bodies, destitute of cleavage or other marks of crystalline structure; and corresponds to the term colloid, subsequently devised by Graham. Among porodic species Breithaupt included, besides colloidal sulphates, phosphates, and arsenates, such bodies as opal, serpentine, chrysocolla, allophane, tachylite,

of which — in many cases, at least — include crystalline portions belonging to sparry types. The tribal distinctions above indicated are farther noticed in § 32.

A similar relation between hardness and chemical indifference and the degree of condensation has also been pointed out by the author for the anhydrous oxyds other than silicates, and for metals and their compounds with sulphur, selenium, tellurium, arsenic, and antimony, which make up the several tribes of the sub-orders Metallometallate and Spathometallate in the great natural order of the Metallates. (16.)

§ 31. The question of the relations of chemical indifference to condensation has lately received a new and important illustration from the study of the silicates. That, while many of these are readily attacked by fluorhydric acid, some few of them resist more or less completely its action, has long been known.

and obsidian — substances alike of aqueous and of igneous origin. Besides an order, *Porodini*, he subsequently defined porodic genera in other orders. See, farther, the author's Mineral Physiology and Physiography, p. 383.

Among the latter had been noticed staurolite and zircon, and, more recently, amphibole, pyroxene, and chrysolite,—a fact of which Fouqué availed himself to separate the last two species from various feldspars and from vitreous silicates. Mr. J. B. Mackintosh, late of Columbia College, New York, and now of Lehigh University, in extending these observations, found, as he informed me, a few months since, that garnet also is indifferent to the action of the acid; a fact ascribed by me to its great condensation, which I had already concluded to be the cause of the similar indifference of the species above named. In confirmation of this view I suggested a comparison between the more condensed species epidote and spodumene, on the one hand, and the less condensed iolite and petalite, on the other; predicting the insolubility of the first two and the solubility of the last two in fluorhydric acid. This prevision was at once confirmed by Mackintosh, who has since greatly extended, and made quantitative, similar experiments on many of the more important silicates.

§ 32. In an attempt at A Natural System in

Mineralogy, in 1885 **(16)**, there was proposed, as already noticed, for the three sub-orders into which the order of Silicates was divided, a classification into tribes, based in great part on the different degrees of condensation; and it may be said, in a few words, that the results of Mackintosh appear in all cases to verify the law of chemical indifference laid down in 1863. While the Pectolitoids, including pectolite, apophyllite, and calamine; the Protospathoids, like willemite and wollastonite; the Zeolitoids; and the various Protoperspathoids, including the feldspars, scapolites, leucite, iolite, and petalite, are more or less completely attacked and dissolved by fluorhydric acid, the Protadamantoids, like pyroxene, amphibole, and chrysolite, resist more or less completely its action. To this tribe, also, rather than to the Pectolitoids, datolite, from its great condensation and its comparative indifference to the acid solvent, appears to belong. The same indifference is observed in the Protoperadamantoids, including garnet, epidote, zoisite, axinite, spodumene, staurolite, and the tourmalines, besides idocrase and prehnite. In like manner, the Peradamantoids andalusite, topaz, and cyanite resist its action. Among the phylloid or micaceous silicates, which are, for the most part, though less hard than spathoids, highly condensed species, the Protophylloid talc, and the Protoperphylloids ripidolite, margarite, zinnwaldite, and muscovite resist the action of the acid; while biotite, phlogopite, and jefferisite are but slightly attacked.

§ 33. The rate of attack under similar conditions varies greatly for different species. When a given weight of the mineral in grains of a determined size is exposed for an hour to a large excess of a dilute fluorhydric acid which fails to attack sensibly the more resisting silicates, Mackintosh finds that, in round numbers, from one to two per cent of amphibole, about five of chrysolite, from twenty-five to thirty-five of petalite, and the feldspars, albite, oligoclase, and labradorite, forty-three of orthoclase, and sixty-six of leucite are dissolved; while willemite, like the colloid species halloysite, is completely dissolved. The same is true of the colloid, opal, while artificial tridymite is also readily attacked, but quartz is more resisting than chrysolite. In experiments on a blast-furnace slag, it was found that portions of the mass, which, on slow cooling, had passed from a vitreous or porodic condition to one of crystallinity, with a notable increase in density, had become much less soluble in the acid. The publication of the farther results of these studies, which are still in progress, and which Mackintosh has kindly communicated to me, will constitute a most important contribution to chemical mineralogy,[1] and a farther proof of the dependence of chemical indifference on condensation.

[1] They have as yet been made known only by preliminary notices, — in the School of Mines Quarterly, in July, 1886; in a paper by me, as yet unpublished, read to the Royal Society of Canada, May 26, 1886; and in an appendix to my recently published Mineral Physiology and Physiography. A more complete account, entitled "The Action of Hydrofluoric Acid on Silica and the Silicates," was read by Mackintosh before the American Chemical Society, in New York, November 5, 1886, and appears in the new Journal of Analytical Chemistry for January, 1887.

# CHAPTER VIII.

## THE ATOMIC HYPOTHESIS.

§ 34. FROM an early time in the history of thought two distinct and opposite conceptions of the intimate nature of matter have had their partisans. The Pythagoreans taught the infinite divisibility of matter, and its continuity in a given mass,—a view held by Plato and Aristotle, and by Kant among modern masters. The opposed doctrine of the finite divisibility of matter, and the existence of ultimate material particles, or atoms, taught by Leucippus, can be traced to Phœnician, Hindoo, and, probably, to Egyptian sources. It was maintained by Epicurus and his school, was lauded by Francis Bacon, and adopted by Newton, though controverted by Descartes, Leibnitz, and Euler; and had fallen into neglect until resuscitated in this century by Dalton, in his System of Chemical Philosophy, published in 1808; since which time it has been accepted by most of the popular writers on chemistry as a necessary part of the law of definite and multiple proportions then unfolded by Dalton.[1] The kinetic hypothesis of the constitution of gases, which conceives them to consist of solid, perfectly elastic spherical particles, moving in all directions, and actuated with different degrees of velocity for different gases, first put forward by D. Bernoulli, in 1747, and adopted by many modern

[1] See, for a history of the atomic hypothesis, Daubeny, Introduction to the Atomic Theory, and, more concisely, Odling, Watts's Dictionary of Chemistry, under "Atomic Weights." Also for an extended discussion of the history of the atomic hypothesis, and the arguments for and against it, Whewell, History of the Inductive Sciences, vol. i. book vi. chap. 5. He there declares that its doctrines, when they "are tried upon the general range of chemical observation, prove incapable of even expressing, without self-contradiction, the laws of phenomena." "Chemical facts not only do not prove the atomic theory as a physical truth, but they are not, according to any modification yet devised of the theory, reconcilable with its scheme." He adds that "when we would assert this theory, not as a convenient hypothesis for the expression or calculation of the laws of nature, but as a philosophical truth respecting the constitution of the universe, we find ourselves checked by difficulties of reasoning which we cannot overcome, as well as by conflicting phenomena which we cannot reconcile."

physicists, was, according to Graham, resuscitated in our own time by Herepath, in 1847.[1]

§ 35. In discussing, in 1853, the nature of the chemical process, as already set forth in Chapter II., it was asserted that, "as chemical combination is not a putting together of molecules, but an interpenetration of masses, the application of the atomic hypothesis to explain the law of definite proportions is wholly unnecessary." (4.) Again, it was said that, since "the volumes of the uniting species are always merged in that of the new one . . . the atomic theory, as applied by Dalton, which makes combination consist in juxtaposition, is untenable." (5.)

§ 36. "If, then, as maintained by the writer, the law of volumes is universal, and if the production of liquids and solids by the condensation of vapors is a process of chemical union or integration, giving rise to polymers the equivalent weights of which are as much more elevated as their densities are greater than those of the vapors which combine to

---

[1] Graham, Chemical and Physical Researches, p. 211, note. For a critical examination of this hypothesis, see Stallo, The Concepts and Theories of Modern Physics, chapters iv. vii. viii.

## The Atomic Hypothesis. 63

form them, the application of the hypothesis of atoms and molecules to explain the law of definite proportions and the chemical process is not only unnecessary but misleading. According to this hypothesis, which supposes molecules to be built up of atoms, and masses of molecules, the different ratios in unlike species between the combining weight of the chemical unit, or molecule (as deduced from analysis and from vapor-density, $H = 1.0$), and the specific gravity of the mass are supposed to represent the relative dimensions of the molecules. Hence, the values got for solids and liquids by dividing these combining weights by the specific gravity have been called 'molecular volumes.' The number of such 'chemical molecules,' required to build up a 'physical molecule' of constant volume would, according to this hypothesis, be inversely as their size. If, however, as all the phenomena of chemistry show, the formation of higher and more complex species is by condensation or integration, or, in other words, by identification of volume, and not by juxtaposition, it follows that the so-called molecular volumes are really numbers representing the relative amounts of contraction of the respective substances in passing from the gaseous to the liquid or solid state, and are the reciprocals of the coefficient of condensation of the assumed chemical units," or atoms, or molecules.[1] (**17.**)

§ 37. Hence it was, that in discussing, in 1885, the relations of density to equivalent, while we admitted the language of the atomic hypothesis, and, in making use of the well known formula $p \div d = v$, recognized chemical units or molecules, whose equivalent or combining weights are represented by $p$, we were careful to say that differences in the density of solid species "are not dependent on variations in the hypothetical units adopted for convenience in calculation, but belong to the species as an integer, and correspond to a greater or less condensation of its mass; that is to say, to the identification in a constant volume of a greater or less number of these chemical units. The very terms of atom and molecule which

---

[1] The paper here quoted, on "The Law of Volumes in Chemistry," is reprinted in the Chemical News for October 26, 1886.

we apply to these imaginary units, and to the mass, are concessions to a popular terminology, and are not only inadequate but, to a certain extent, misleading when applied to chemical operations." (**16**.)

§ 38. The confusion which, as already pointed out (§ 11), exists in the minds of many chemists between dynamical and chemical activities, and prevents a clear conception of the nature of the chemical process, has led to the transference of the atomic or molecular hypothesis from the theory of dynamics to the theory of chemistry. The phenomena of elasticity, of the movements of gases and liquids, of temperature, of electricity, and of radiant energy, — in a word, all the manifestations which come under the head of dynamics, — are, in the opinion of many of their students, most easily explained if we suppose that the species which is the subject of these phenomena has a structure, not continuous, but made up of discrete molecules or atoms of a definite and constant size, which in liquid and solid bodies varies for different species.

§ 39. The acceptance of this hypothesis of the constitution of matter for all species, whether solid, liquid, or gaseous, will be seen, on reflection, to have no direct bearing on chemism; which does not consider the species as such, but only its relations to the species from which it has been derived, or into which it may pass, by processes of specific integration or disintegration, either homogeneous or heterogeneous.

If it were possible to demonstrate the existence of a molecular structure, for example, in a crystal of calcite; to fix, as has been attempted, by calculation, the dimensions and the weight of the constituent molecules by the juxtaposition of which the crystal is built up, we should have no better warrant than before for the hypothesis that, in these molecules of calcite, either the unlike chemical species of carbonic dioxyd and of calcic oxyd, or those others, carbon, calcium, and oxygen, into which chemical disintegration can resolve the calcite, are present as such. "Of the relations which subsist between the higher species and those derived from them we can only assert the possibility, and, under certain conditions, the certainty of producing the one from the other." (4.)

In considering this subject in 1874 it was

said: "Are we not going beyond the limits of a sound philosophy when we endeavor by hypotheses of hard particles with void spaces, of atoms and molecules with bonds and links, to explain chemical affinities; and when we give a concrete form to our mechanical conceptions of the great laws of definite and multiple proportions to which the chemical process is subordinated? Let us not confound the image with the thing itself, until, in the language of Brodie, in the discussion of this very question, 'we mistake the suggestions of fancy for the reality of nature, and cease to distinguish between conjecture and fact.'. The atomic hypothesis, by the aid of which Dalton sought to explain his great generalizations, has done good service in chemistry, as the Newtonian theory of light did in optics, but is already losing its hold on many advanced thinkers in our science." **(12.)**[1]

[1] The reader may consult with much advantage, in this connection, a paper by C. R. A. Wright, On the Relations between the Atomic Hypothesis and the Condensed Symbolic Expressions of Chemical Facts and Changes known as Dissected (Structural) Formulæ, in 1872, L., E., and D. Philos. Mag. (4) xliii. 241-264. See, also, for R. A. Atkinson's strictures thereon, and Dr. Wright's rejoinder, *ibid.*, 428-433, 503-514.

# CHAPTER IX.

### THE LAW OF VOLUMES.

§ 40. "THE quantitative relation of one mineral (chemical) species to another is its equivalent weight." (**10.**) As regards the relation of weight to volume, it was said, in 1853: "The weights of equal volumes of gases and vapors are their equivalent weights; and the doctrine of chemical equivalents is that of the equivalency of volumes. According to the atomic hypothesis, these weights represent the relative weights of the atoms; and, as equal volumes contain the same number of atoms, these must have similar volumes, so that we come at last to the equivalency of volumes." The condensation which constitutes polymerism "evidently offers no exception to the law of equivalent volumes." (**4.**) Dalton, in his remarkable generalizations, which are summed up in his theory of definite and multiple proportions, "linked his discoveries with the old hypothesis of the atomic constitution of matter, which is, however, by no means necessarily connected with the great laws of combination by weight and by number. It was reserved for Gay Lussac [1] to make known a not less beautiful generalization, by showing that in the combination of vapors and gases there exists an equally simple relation of volumes, and that measure, not less than number and weight, governs all chemical changes." "All things, declares the sage,[2] are ordered by measure, by number, and by weight." (**12.**) Already, in 1853, it had been said: "The simple relations of volume, which Gay Lussac pointed out in the chemical changes of gases, apply to all liquid and solid species, thus leading the way to a correct

[1] In his Theory of Volumes, Mémoires d'Arceuil, 1809, ii. 207.

[2] This is, of course, a free reading of the passage which we have chosen for the motto of this volume: "Omnia mensura, et numero, et pondere disposuisti," Liber Sapientiae, cap. xi., translated in the received English version of the Bible: "Thou hast ordered all things in measure, and number, and weight," Wisdom, xi. 20. The Greek original of this is made by Daubeny the epigraph to his Introduction to the Atomic Theory.

understanding of the equivalent volumes of the latter." (**5.**)

§ 41. We had thus in the development of a theory of chemistry clearly defined the principle that the doctrine of chemical equivalents is that of the equivalency of volumes; and that the law of volumes is universal, and applies not only to gases but to liquids and solids, which are species distinct from their corresponding vapors. We had, moreover, already, in 1853, inquired "whether a definite and constant relation may not be discovered between its vapor-density and the specific gravity of a species in its solid state," and had asked "what equivalent corresponds to a given specific gravity in any crystalline solid; or, in other words, what is the value of the condensation which takes place in the change from the gaseous to the solid state?" It was farther said: "Such a relation being established, and the value of the condensation in passing from a gaseous to a solid state being known, the equivalents of solids, like those of gases, might be determined from their specific gravities." (**4.**) It remains to be shown why the solution of the problem, now offered for the first time in 1886, which seems an obvious deduction from the principles thus laid down in 1853, was not sooner discovered.

§ 42. The explanation is to be found in the fact that, while affirming in a broader sense than had before been stated the universality of the law of volumes, the writer still believed, in accordance with the generally received and as yet unquestioned teaching of those who had studied the volumetric relations of solid species, that this law was conditioned in some manner by crystalline form, so that the volume of solids (and of liquids, also) was an arbitrary and a variable quantity, instead of being, as in the case of gases and vapors, an ideal unit. The general principle laid down by Gmelin, in his Handbook of Chemistry, as deduced from the studies of the volumes of solids, is that "isomorphous substances have equal atomic volumes."[1] Otto, also, in his learned review of the subject, declares that Dumas, whose studies, in connection with Leroyer, have been the point of departure for all investigations of the

[1] Gmelin, Cavendish Society's edition, 1848, vol i.

relation between the specific gravity and the chemical composition of solids, discovered that the relation of equal volumes "is connected with the crystalline form, as it exists only in isomorphous species." He adds, in regard to the formula $p \div d = v$, "Dumas was the first to demonstrate that the value of $v$ was sensibly the same for all bodies whose isomorphous relations had been established by Mitscherlich."[1] Following the line thus indicated, chemists have been led to compare chiefly such isomorphous groups, obtaining results which have served to confirm the belief in the correctness of this conclusion as to the significance of crystalline form.

§ 43. All this was before me in 1853, when I wrote, with regard to "some allied and isomorphous species," that "H. Kopp, in dividing the assumed equivalent weights of such bodies by their specific gravities, obtained quantities which were found to be equal for some of these related species. These numbers evidently represent the volumes of equivalents, and, in accordance

[1] Otto, Chemical Reports and Memoirs of the Cavendish Society, 1848, p. 67.

with the atomic hypothesis, are said to denote the atomic volumes." As regards the published investigations up to that time, it was then said: "Their results show that the volumes thus calculated for related species of similar crystalline form are generally identical, or sustain to each other some simple ratio." From the comparisons of the values of $v$ in the alums and in the hydrated sodium-orthophosphates and ortharsenates, and from other examples which were apparently in accordance with the teachings of Dumas as to the identity of volume in related isomorphous species, the conclusion was then reached that "what are called the atomic volumes of crystallized species are the comparative volumes of their crystals" (4); thus stating in other terms the conclusion of Dumas that, while isomorphous substances have identical volumes, the volumes in different crystalline systems were unlike; a point of view which was then illustrated at some length. In 1855, I had more than one conference on this problem with Dumas, who had read with approval, and then discussed at length, my three papers of 1853 and 1854, while he ingeniously explained the apparent exceptions to his law indicated in the above extracts, by the principle of polymerism.

§ 44. That some broader principle than that involved in isomorphism underlies these apparent conformities in volume was soon after perceived, though but imperfectly apprehended; and when, in 1874, the essay above quoted was republished in the writer's volume of "Chemical and Geological Essays," a foot-note was added to the portion from which was taken the above cited passage as to the relation of volume to crystalline form, saying, " The conclusions in this paragraph may be liable to correction; but I leave them as they were printed twenty-one years since." It will be noted that, in reviewing this subject in 1867, it was said: "The variable relation to space of the empirical equivalent of non-gaseous species or, in other words, the varying equivalent volume . . . shows that there exist in different species very unlike degrees of condensation" (10); no reference being made to crystalline form as in any way conditioning this condensation. The teaching of Dumas and his successors in this field of inquiry for some years, however, maintained the writer, in common with other chemists, in the belief that the accident of crystalline form has some intimate relation with condensation. That the similar values of $v$ in the isomorphous species compared involve a relation of equivalents not then suspected, is made apparent by the results of Graham's studies in liquid diffusion, then made known, but not fully understood even by himself; to the consideration of which we shall return in Chapter XII.

§ 45. When we have once attained the conception that the law of volumes is a fundamental and a universal law, to which all species, whether gaseous, liquid, or solid, whether colloid or crystalline, and all changes of state among these, are subordinated, the solution of the great problem proposed in 1853, and stated in § 41, is seen in the familiar processes of the conversion of water into steam and of steam into water. The latter is the change of the vapor $H_2O$ into that polymeric liquid species which, at its maximum density, is the unit of specific gravity for all liquids and solids. According to the calculation in Ganot's Eléments de Physique,[1] based on a comparison of the densities of steam as compared with air, and of air as compared with water, the ratio between the weights of equal volumes of steam at 100° and 760 mm. pressure, and of water at 0° is 1 : 1698. The weights of equal volumes of air and water, both at 0°, are as 1 : 773; but those of air at 100° and water at 0° as 0.73178 : 773. The density of steam at 100° and 760 mm. as compared with air under the same conditions being as 0.6225 : 1, it follows that the ratio between the weights of equal volumes of steam at 100° and water at 0° is as 0.73178 × 0.6225 : 773 = 0.4555 : 773 = 1 : 1698.

§ 46. But we may approach this question more directly, by calculating the specific gravity of steam from that of hydrogen. Regnault found for the weight of one litre of hydrogen in the latitude of Paris at 0° and 760 mm. pressure, 0.089578 gramme, which from its expansion as determined by him would at 100° be 0.065572 gramme. If we assume the equivalent of oxygen as 16.0, we have for the weight of a litre of steam, $H_2O$, at 100° and 760 mm., 0.590148

[1] Ganot, Atkinson's translation, fifth edition, 1872, p. 290.

gramme, and, dividing by this number the weight of one litre of water at $4° = 1000.0$ grammes, we get 1694.49. If, however, we take for oxygen the number 15.96 (being that deduced by Stas), we have for the weight of a litre of steam at 100° and 760 mm., 0.58894 gramme, and, taking this number as the divisor, obtain 1697.96; but, as water at 0° has a specific gravity of 0.99987 (Rosetti), the ratio between steam at 100° and 760 mm. and water at 0° becomes 1 : 1697.74.

The close approximation to the directly calculated ratio 1 : 1698 got by employing the corrected equivalent weight of oxygen is such that we may assume this figure as representing the true ratio. But, in order to establish the amount of condensation in the conversion of steam into water, we must take the specific gravity of this liquid at the temperature of condensation, 100°; and this, according to the determination of Kopp, is 0.95878.

$$1.00000 : 0.95878 :: 1698 : x = 1628.04.$$

This figure represents the number of volumes of steam at 100° and 760 mm. which are condensed in one volume of water at the same temperature. But the weight of the litre of hydrogen gas at the standard temperature and pressure of 0° and 760 mm. $= 0.089578$ grams, being the basis of calculation in equivalent weights, that of the litre of water-vapor, also calculated for standard temperature and pressure, $H_2O = 17.96$, is 0.8044 grams. Comparing this with the weight of a litre of water at 100° (the temperature of its formation at 760 mm.), which $= 958.78$ grams, we have:—

$$0.8044 : 958.78 :: 1 : x = 1191.9.$$

Water thus contains condensed in one volume approximately 1192 volumes of water-vapor, which makes its equivalent or integral weight $1192(H_2O) = 21,408$. Calculating directly from the weight of hydrogen gas as given above, we have:—

$$0.089578 : 958.78 :: 2 : x = 21,406.6.$$

In the uncertainty which still prevails as to the exact weight of oxygen, hydrogen being unity, we may safely assume in the calculation of densities of other liquid and solid species, the number 21,400 as a sufficiently close approximation to the integral weight of water, the density of which has been made the unit of specific gravity for all such species. Ice is perhaps $1094(H_2O) = 19,648$; which corresponds to a specific gravity of 0.9181.

In 1885, when the problem of fixing the equivalent weights of liquid and solid species had not yet been solved, it was shown that this weight for certain solid species, such as the salts of the cobaltic ammonias, and the polytungstates, must be several thousand times that of hydrogen (§ 26), with correspondingly large equivalent volumes. Having then noticed what, in the language of the atomic theory, were called the unit-weight and the unit-volume (§ 48), represented by P and V ($=p$ and $v$), it was said: "The relations alike of this unit-weight and unit-volume to those of the molecule to which it belongs are unknown. But this molecule has, by our hypothesis, a constant volume, for which an expression is yet wanting, and can, so far as known, only be attained by assuming as unity the number which corresponds to the highest discovered value of V. The true unit of molecular volume will probably still be some multiple of this number." (**16.**)

# CHAPTER X.

### METAMORPHOSIS IN CHEMISTRY.

§ 47. CHEMISM may be comprehensively defined as the production of new chemical species, either by integration or by disintegration; and these changes, as has been farther shown, may be either homogeneous or heterogeneous. In heterogeneous integration two species of unlike centesimal composition unite; and in heterogeneous differentiation a species is resolved into two or more unlike ones. This genesis of complex species having a centesimal composition unlike the parents, we have designated as *metagenesis*, as set forth at some length in Chapter II. In homogeneous chemical change we have the production of new species by the union or integration of two or more like species, either elemental, or, if complex, of identical centesimal composition; or conversely, the disintegration of such compounded species into simpler and like forms. These homogeneous chemical changes constitute what we have distinguished as *metamorphosis* in chemistry, and include both polymerization and depolymerization. They have long been familiar to chemists, alike in so-called elements and in many hydrocarbonaceous bodies, as has been noticed in § 5, but their importance is such as to demand a special discussion.

§ 48. It has already been shown that the bodies derived from gases and vapors by liquefaction and solidification are distinct species, and are polymers of these; their generation being by a process of homogeneous integration, or, in other words, of metamorphosis by condensation; while the vaporization of such liquids and solids is homogeneous disintegration, or metamorphosis by expansion. For all species which are said to volatilize without decomposition, — that is to say, which may be converted into vapor without heterogeneous differentiation; or which, in other words, yield thereby simpler species having the same centesimal composition as their parent, — the equivalent weights are known from their vapor-densities; and the coefficient of condensation for such liquid and solid species is directly found by comparing the density of these gaseous species with that of their liquid and solid polymers, as already shown in the case of steam, water, and ice.

When, however, we have to deal with species which are too fixed in the fire to admit of vapor-determinations, or with those which by heat undergo heterogeneous disintegration, — as, for example, sugars, carbon-spars, and hydrous silicates, — in place of the unit-weight deduced from vapor-density (as in the case of water, mercuric chlorid, or sulphur), we may make the simplest formula deduced from analysis serve as the unit; or, for greater convenience in calculation, may, in the case of oxyds and silicates, assign to it a value corresponding to $H = 1$ and $O = 8$. The unit for silica thus becomes $SiO_2 = 60 \div 4 = 15$; that for alumina $Al_2O_6 = 102 \div 6 = 17$; and that for the magnesian silicate forsterite, $SiMg_2O_4 = 140 \div 8 = 17.5$. Such unit-weights have been employed by the writer in his late essay on A Natural System in Mineralogy (**16**), in the tables of which they have been represented by P; while the values got by dividing these numbers by the specific gravities have been called unit-volumes, and designated by V. That the units which make up the polymers in such non-volatile species are, however, far greater than these unit-weights is apparent when we consider the highly complex formulas, and the elevated equivalent weights, to which we are led by the analyses of many species, as already noted in Chapter VI.

§ 49. To similar conclusions are we conducted by the study of the phenomena of metamorphosis in certain hydrocarbonaceous species, notable examples of which are seen in the aldehydes, and in the turpentine-oils. It will be remembered that normal acetic aldehyde, with a vapor-density corresponding to $C_2H_4O$ and a specific gravity of about .800, boiling at 21°, is, in presence of small quantities of various reagents, such as chlorhydric acid or sulphurous anhydrid, at ordinary temperatures, rapidly converted, with considerable evolution of heat, into paraldehyde, a liquid of specific gravity .998, boiling at 124°, becoming a crystalline solid at 10°, and, from its vapor-density, a tri-aldehyde

$3(C_2H_4O)$. The same reagents, at low temperatures, convert aldehyde into another modification, metaldehyde, which is a crystalline solid at 100°, and at a higher temperature volatilizes without fusion. From the readiness with which (like paraldehyde) it is, at a higher temperature, reconverted into normal aldehyde, the vapor-density of this polyaldehyde $x(C_2H_4O)$ cannot be determined.

Chloral, $C_2(HCl_3)O$, in like manner, readily changes into a white insoluble species, metachloral, of unknown equivalent weight, which at 180° is converted into the vapor of normal chloral, volatile at 95°. Still more remarkable is the case of formic aldehyde, or methylene oxyd, at ordinary temperatures a gas soluble in water, and, as shown by its vapor-density, $CH_2O = 30$, which spontaneously changes into a white, insoluble polymer. This, which is volatile at 100°, melts at 152°, and its vapor, which is perhaps $3(CH_2O)$, is at a higher temperature metamorphosed into the normal aldehyde, which polymerizes again on cooling. We have thus an example of two interconvertible species of identical centesimal composition, the one a gas, soluble in water, and the other a white insoluble solid.

§ 50. Chemists have long distinguished among the turpentine-oils a considerable number of species which, while having the same centesimal composition, differ considerably not only in chemical relations, in optical characters, and in being liquid or solid at ordinary temperatures, but in specific gravity and in boiling-point. Some of these are found in nature, while others are produced by the transformations of ordinary French or American turpentine-oil. This, to which the formula $C_{10}H_{16}$ is assigned, is readily metamorphosed by heat, and by many reagents, giving rise to a large number of new species. Some of these belong (together with various natural oils) to the class of proper terpenes, including several groups, to all of which is assigned the above formula. Among these natural and artificial terpenes are the liquid pinenes, boiling at about 160°; the camphenes, having a similar boiling-point, but forming crystalline solids below 50°; and the limonenes, liquids boiling about 175°-177°, — with several other groups — each of these including varieties or sub-species, differing in optical characters and in chemical relations. Besides these true terpenes are others, regarded as sesquiterpenes ($C_{15}H_{24}$), boiling at 250°–260°; diterpene, or colophene ($C_{20}H_{32}$), boiling at 300°; and polyterpenes, of still higher equivalent and unknown complexity, boiling above 360°.

§ 51. When the vapor of turpentine-oil is passed through a tube heated to low redness, the products of condensation yield, besides different terpenes, and colophene, a considerable portion of a very volatile liquid, boiling between 37° and 40°, with a vapor-density showing it to be a hemiterpene ($C_5H_8$), which is designated pentine, and is apparently identical with what have been called isoprene and valerylene. Besides these homogeneous transformations or metamorphoses, both of integration and differentiation, produced by heat in turpentine-oil, there is a simultaneous heterogeneous differentiation of a portion, resulting in the production of a homologue of pentine, namely, heptine ($C_7H_{12}$), and probably, also, of the intermediate hexine, together with hydrocarbons of the benzene-toluene series, and some permanent gases.

When pentine is heated in sealed tubes to 280°, it undergoes a partial polymerization, a portion of it being changed into a terpene, with a little colophene. When distilled in open vessels, the transformation of this volatile liquid into less volatile products causes the temperature of the mass to rise rapidly, without the aid of heat from without, the action sometimes becoming explosive. A similar rise of temperature is noticed in the rapid polymerization of aldehyde, and in that of ordinary turpentine-oil, which is produced by contact with sulphuric acid or with small portions of fluorid of boron.[1] Wallach, who has lately reviewed the history of the various turpentine-oils, and the products of the polymerization of pentine and terpene, has designated these as pentine ($C_5H_8$); di-pentine, including the various turpenes ($C_{10}H_{16}$); tri-pentine, including also

---

[1] See Watts, Dictionary of Chemistry, under Turpentine-Oil; also the papers of Armstrong and Tilden, Journal of the Chemical Society of London, vols. 35 and 45; and later, Wallach, Liebigs Annalen, vol. ccxxvii., p. 277.

various sub-species ($C_{15}H_{24}$); and tetra-pentine, or colophene ($C_{20}H_{32}$); — besides the higher polymers, of unknown complexity, including the so-called metaterpene. The specific gravity of pentine (mono-pentine) is about .680; that of the terpenes varies, but is not far from .850; and that of the higher polymers is from .920 to above 1.00.

§ 52. The disengagement of heat, noticed in the polymerization of aldehyde, of pentine, and of turpentine-oil, is analogous to that, resulting in vivid incandescence from internal change, which takes place when many amorphous bodies are heated to low redness, by which change, in the language of Gmelin, they acquire "greater specific gravity, greater hardness, and less solubility." This incandescence is observed in titanic, tantalic, molybdous, zirconic, chromic, and ferric oxyds, in magnesium pyrophosphate, and in certain artificial arsenates and antimonates, as well as in native minerals, like euxenite, gadolinite, and allanite. Such as these were called by Scheerer pyrognomic species, and adduced by him as evidence that the granitic veinstones in which they are found were not formed at very elevated temperatures.[1]

§ 53. We have already noticed the question of metamorphosis among so-called elemental species, as discussed in 1848 (§ 5). As regards sulphur, and the threefold condensation of its vapor then suggested, it will be remembered that, as was shown by Bineau, in 1869, this vapor, at temperatures approaching 1000°, is expanded to that of normal sulphur. To this we have a parallel in the cases of paraldehyde and metaldehyde, readily converted into aldehyde, and in turpentine-oil, itself a polymer, which may be volatilized as such, but at an elevated temperature is resolved into pentine ($C_5H_8$), which, as we have seen, readily passes again into $C_{10}H_{16}$, and still higher polymers.

The effect of a given temperature upon a vapor cannot be determined *à priori;* while sulphur and turpentine-oil volatilize as polymers, the vapors of which undergo homogeneous differentiation only at much higher temper-

[1] See, in this connection, Gmelin's Handbook, Cavendish Society's Edition, I. 106, 107; also the author's Mineral Physiology, etc., p. 96.

atures, the vapor of water assumes at once its normal type, $H_2O$, and at higher temperatures undergoes heterogeneous differentiation, or metagenesis, being resolved into hydrogen and oxygen. There are, doubtless, other gaseous polymers, which, like tri-sulphur, $S_3$, and di-pentine, $2(C_5H_8)$, exist within certain limits of temperature and pressure. In this connection, the studies of Cagniard de la Tour, of Drion, and of Andrews, on the conversion of liquids into gases, are very important, and help to enlarge our conceptions of this polymerism in vapors under great pressure; while farther homogeneous integration, at lower temperatures, gives rise to liquid and solid species, more or less stable at the ordinary atmospheric pressure.

§ 54. The same gaseous species may yield two or more liquid or solid species of different degrees of density, and of different stability as well, as is evinced by differences in boiling-point. This is seen from the comparison of the various groups of liquid and solid terpenes, or di-pentines, all of which are distinct from the liquid forms alike of true pentine (mono-pentine), of tri-pentine, and of tetra-pentine, and constitute different liquid polymers of $2(C_5H_8)$. Analogous differences are observed in the optically active and inactive amylic alcohols found together in the products of fermentation, which transmit these differences to their derivatives, and present at the same time differences in their chemical relations. The various di-pentines are to be compared to ice and water. The latter, under favorable conditions, is a liquid of density 1.000 (nearly), even at —10°, though then readily transformed into ice, with a density of 0.917. In like manner, we have, at ordinary temperatures, besides the unstable liquid phosphorus of density 1.77, the ordinary and readily fusible species with a density of 1.82, and the red crystalline species of density 2.35, all of which, so far as known, are directly changed by heat into the same elemental or normal vapor, without gaseous polymerism, precisely as both ice and water pass directly into the normal vapor, $H_2O$.

§ 55. A remarkable example of metamorphosis in an elemental species is afforded by tin, which, as observed by Fritzsche, Oudemans,

Schertel, and Richards,[1] when exposed to great natural cold, is changed from the white, tough, malleable metal, having a specific gravity of about 7.30, to a gray, brittle, crystalline body, of much less density. This change takes place not only in pure block tin but, as observed by Richards, in ingots containing 2.5 per cent of mercury, the change, at a minimum temperature of about — 18°, gradually extending, in this case, from certain centres, so as in a few weeks to involve the whole mass; which had a radiated structure, and was compared in aspect to stibnite. Its density was found by Richards to be 6.175, that of the unchanged portions being 7.387. Other observers have found for this altered form, with pure tin, densities of 6.0 and even of 5.8. Its instability makes this determination difficult, since the gray, brittle species is readily changed into white tin by strong pressure, by a sudden blow, or by immersion in hot water, which restores to it its toughness and its original density of 7.30. The

---

[1] Watts's Dictionary of Chemistry, Third Supplement; also Richards, in Transactions American Institute of Mining Engineers, xi. 221.

gray tin exhibits, alike in acid and in alkaline liquids, electrical relations different from those of the ordinary metal.

This metamorphosis of tin is comparable to that of water into ice by cold, or that of aragonite into calcite by heat, the transformation in each case being one of homogeneous disintegration from a denser to a lighter species. It is not improbable that similar changes of state analogous to this may take place in other bodies, and that the temporary production of a similarly brittle and chemically unstable form of iron, under like circumstances, may explain the apparently ready frangibility of that metal when exposed to severe frosts.

§ 56. In 1848 (§ 5), in calling attention to the fact that sulphur-vapor in the only form then known was tri-sulphur, SSS, which was compared to SOO, it was suggested that ozone might be tri-oxygen, OOO. It however is now known that, though condensed oxygen, its equivalent weight corresponds to one-half of this, or to OOO doubled in volume by homogeneous disintegration. Making $H_2$ the unit of volume, we have for the vapor of tri-sulphur

$S_6$, and for sulphur-vapor above 500° $S_2$. If, then, normal oxygen is $O_2$, ozone is not $O_6$, but $O_3$. So, the vapor of iodine, $I_2$ at temperatures approaching 1500°, is, as shown by Crafts, changed in great part into a species, $I_1$, having but one-half the density which belongs to normal iodine-vapor. A similar change, to a less extent, is wrought at high temperatures in bromine and chlorine, as shown by Victor Meyer and others. These so-called elemental gaseous species thus undergo at higher temperatures a homogeneous disintegration, like that of the vapor of di-pentine, when, by heat, it is changed into pentine.

§ 57. All of these, and similar changes, are but examples of that law of dissociation by heat, which includes alike so-called elements and compounds; and, as the writer was the first to teach, in 1867, is universal, and applies to the chemistry of stellar matter. While oxygen, nitrogen, and hydrogen give no evidence of similar dissociation in our laboratories, it was then suggested that they may suffer it in the solar fires; and later, in 1874, it was pointed out that the unknown element, appar-

## Metamorphosis in Chemistry. 95

ently of great tenuity, shown by the spectroscope in the solar chromosphere, as line 1474, Kirchhoff, is probably a more elemental form of matter than any known on earth, if not the primal element from which all others have been generated, as set forth at length in Chapter III. It is not improbable that the power of flame or the electric spark to effect the sudden union of chlorine and of oxygen with hydrogen may be due to the effect of intense heat in separating momentarily into simpler forms portions of these gases, so that we may have in the process of their combination a union of unknown elemental and less dense species.

§ 58. Whether in any case the homogeneous differentiation or depolymerization of a volatile condensed species will be effected at a single step as, under the ordinary pressure at least, is the case with water, the alcohols, and most other liquids; or whether, as with sulphur or iodine, or with the gaseous polymers of pentine and aldehyde, by two or more successive steps, cannot, in the present state of our knowledge, be foreseen. When, therefore, we have to do with species which are not susceptible of metamorphosis by expansion, — that is to say, of integral volatilization, — we cannot, in studying their condensation, know whether we have to deal with the simplest possible formula, or with some multiple thereof. Did we not know the existence alike of mono-pentine, and of its polymers, di-pentine, tri-pentine, and tetra-pentine, we could not tell how to write the formula of solid camphene, or of liquid isoprene, pinene, limonene, or colophene, although the equivalent weights of these had already been fixed from their specific gravities. In other words, we could not determine whether the unit by the polymerization of which these liquid and solid species have been generated is, in any given case, $C_5H_8$, $C_{10}H_{16}$, $C_{15}H_{24}$, or $C_{20}H_{32}$. In like manner, we might ask whether the unit in crystalline sulphur is mono-sulphur, $S = 32$, or tri-sulphur, $S_3 = 96$. Yet the determination of these numbers, in the one case and in the other, is necessary in order to fix the coefficient of condensation in liquid and solid species.

§ 59. Resuming here what has been said of metamorphosis, and its relations to metagenesis, we repeat, that the systems of structural formulas, and of types in chemistry, are but expressions of the genealogical relations of species, as deduced from a consideration of the phenomena of metagenesis. This precedes metamorphosis, which, by homogeneous integration, converts the simpler or normal species into polymeric species of higher equivalent weights. In these condensed species, as already pointed out, there is observed an increase, corresponding to the degree of condensation, in their power of resistance to chemical change; as shown in diminished volatility, fusibility, and solubility, not less than in increased hardness. Carbon dioxyd, $CO_2$, is a gas, and only under exceptional conditions gives rise to liquid and solid polymers, which are rapidly depolymerized at the ordinary pressure and temperature; while the corresponding silicon dioxyd, $SiO_2$, is scarcely known except in insoluble crystalline or colloidal polymeric forms, although it is probable that, at a very high temperature, it assumes the gaseous form of normal silicon dioxyd.[1]

[1] In an experiment made by the writer in September, 1885, with the electrical furnace of the Messrs. Cowles, in which the

§ 60. Many complex species, when generated by double decomposition in aqueous solutions, remain dissolved for a greater or less time before separating in an insoluble condition. The interval which thus elapses before the formation of such precipitates, as in the case of ammonio-magnesium phosphate, of strontium sulphate, and of gypsum, among others, is that required to convert the soluble normal species, by homogeneous integration, into insoluble polymeric species.

Theoretically, all species may exist in their normal or soluble forms; and it is this condition which constitutes the nascent state of bodies. The manner in which oxyds like those of silicon, aluminium, and iron are not only found in nature, but are separated in our laboratories in a crystalline state from aqueous solu-

current of a dynamo-electric machine, of thirty horse-power, was passed through a horizontal column of fragments of charcoal, mingled with pure silicious sand, while a portion of this was reduced to silicon, and a still larger amount fused into a glass, a small portion was found in botryoidal masses, like chalcedony in aspect, adhering to the covering tiles of the furnace, apparently from direct volatilization. See Transactions American Institute of Mining Engineers for 1885; and Chemical News for Nov. 6, 1885; pp. 235-236.

tions, shows that these oxyds, before passing into insoluble and highly condensed polymeric forms, have, like the compounds named above, enjoyed a temporary solubility in water. The very soluble forms of silicic, titanic, stannic, tungstic, ferric, chromic, and aluminic oxyds, which were especially studied by Graham, are well known examples of this. That the mineral silicates, sulphids, etc., found in veinstones have also at one time been in similar conditions, is evident from geognostical observations.[1] This question was discussed by the present writer in 1872, when it was said that "it will probably one day be shown that for the greater number of those oxygenized compounds which we call insoluble, there exists a modification soluble in water." The tendency of most normal species to pass into polymeric or condensed forms, of greater or less stability, is a fact of fundamental importance in chemistry.

[1] Chemical and Geological Essays, page 223, *note*.

# CHAPTER XI.

### THE LAW OF DENSITIES.

§ 61. THE significance of the question raised in § 58 of the last chapter will be made more apparent in considering the densities of some well known solid species; as, for example, the minerals included under the name of calcareous spar or calcite, and consisting of calcium-carbonate. In calculating, in Chapter IX., the equivalent weight of water, we have taken for oxygen the value 15.96, as a closer approximation to its equivalent weight, hydrogen being 1.0, than the generally received number, 16. This correction makes for water, 1192($H_2O$), into which oxygen enters so largely, a noteworthy difference, giving instead of 21,456 the equivalent or integral weight of 21,408; and for this, in calculating the equivalent weights of liquid and solid species, for which water is taken as the unit of specific gravity, we have, for reasons already given (§ 46), proposed to adopt the number 21,400. In species into which oxygen enters in smaller proportion than in water, the difference made by its corrected equivalent weight is of less moment. Thus, for calcium carbonate, $CCaO_3$, which with $O = 16$ has a combining number of 100, we find with $O = 15.96$, the number 99.88.

§ 62. Assuming 2.730 as the specific gravity of calcite (water $= 1.000$), and taking the integral weight of water at 21,400, we have for that of calcite itself:—

$1.000 : 2.730 :: 21,400 : x = 58,422.$

Dividing the integral weight thus deduced by the first combining number — 100 — we have as the coefficient of condensation 584.22, or, rejecting the fraction, 584($CCaO_3$). Substituting now the second number — 99.88 — we get instead an integral weight of 58,330. But since the specific gravity is directly as the integral weight we have for these two values, respectively, the following numbers:—

$21,400 : 58,400 :: 1 : x = 2.7290.$
$21,400 : 58,330 :: 1 : x = 2.7257.$

The small difference of .0033 in these two calculations of the specific gravity of calcite, which includes nearly one-half its weight of oxygen, being within the limits of error in the ordinary determinations of density in solid species, may be disregarded, and in studying the density of the various native oxyds and silicates, the equivalent weights as generally calculated, with $O = 16$, may be employed.

§ 63. Now, while the density of any given gas or vapor is constant at a given temperature and pressure, that of many liquids and solids of identical centesimal composition offers greater or less variations, as may be seen by comparing the specific gravities of the different forms of phosphorus, of tin, of the various acetic aldehydes, of the turpenes, of ice and water, of quartz and tridymite, of aragonite and calcite, with one another, and even of different examples of calcite among themselves. These differences of density among related species, upon which the writer has constantly insisted, but which have not hitherto received the attention that belongs to them, are evidently connected with the greater or less condensation of matter in the species, or, in other words, with polymerization. Until, however, we had attained to a clear conception of the greatness of the condensation, and the consequent high equivalent weight of such species, and of its relation to their respective specific gravities; and, in accordance with what may perhaps be called "the law of densities," could determine the weights of liquids and solids as compared with water, it was impossible to give to these variations in specific gravity their true significance.

§ 64. It is now nearly half a century since Breithaupt, by greatly multiplied and most careful and delicate observations of the genus Carbonites, under which he included, with other rhombohedral carbonates, the true calcites or calcium carbonates, distinguished, among these, numerous species and sub-species, marked not only by variations in angular measurements, in lustre, cleavage, and other superficial characters, but in hardness and in specific gravity. Referring the student for details to the second volume of Breithaupt's Handbuch der Mineralogie (1841), it will be sufficient to note the names and the densities of the species and sub-species of rhombohedral calcium-carbonate thus defined : —

Carbonites archigonius
 Sub-species C. a. levis . . . . 2.690–2.710
  "   C. a. ponderosus . . 2.734–2.754
Carbonites paroicus . . . . . . . . 2.652–2.678
Carbonites eugnosticus
 Sub-species C. e. epithematicus . 2.700–2.708
  "   C. e. mediocris . . . 2.716–2.720
  "   C. e. hypothematicus, 2.720–2.730
Carbonites diamesus
 Sub-species C. d. polymorphus . 2.707–2.714
  "   C. d. mediocris . . . 2.721–2.727
  "   C. d. syngeneticus . 2.732–2.749
Carbonites meroxenus . . . . . . . 2.689–2.705
Carbonites haplotypicus . . . . . . 2.728–2.729
Carbonites melleus . . . . . . . . 2.695–2.697

Of these, *C. diamesus* is by far the most abundant and widely diffused species. The hardness of these calcites varies from $4\frac{1}{2}$ and 4 to $3\frac{3}{4}$ on the scale of Breithaupt, the heavier being the harder species; while the range of densities observed is from 2.652 to 2.754.

§ 65. Reverting to the law of densities already set forth, and taking $CCaO_3 = 99.88$, we arrive, for calcites of different coefficients of condensation, and for aragonite, at the densities given on the following page.

It is evident that, while the *Carbonites paroicus* of Breithaupt corresponds nearly in density to 572($CCaO_3$) or 576($CCaO_3$), *Carbonites archigonius*, sub-species *a. ponderosus*, attains the density of 588($CCaO_3$), the other calcites corresponding to some intermediate term; while that of 630 ($CCaO_3$) agrees closely with the known density of aragonite, which was referred by Breithaupt to another genus, and according to him includes *Holoëdrites haplotypicus* and *H. alloprismaticus*.

|  |  |
|---|---|
| 572($CCaO_3$) | 2.670 |
| 576($CCaO_3$) | 2.688 |
| 580($CCaO_3$) | 2.707 |
| 584($CCaO_3$) | 2.726 |
| 588($CCaO_3$) | 2.744 |
| 630($CCaO_3$) | 2.940 |

§ 66. The question here arises as to the value of the polymeric unit in these calcium carbonates with densities varying from 2.652 to 2.940. In case of acetic aldehyde and the products of its metamorphosis, we are enabled by vapor-densities to show that in one case we have to do with $C_2H_4O$, and in another with $3(C_2H_4O)$; while in the turpentine group we have in liquid pentine a condensation of $C_5H_8$; the various liquid and solid di-pentines of different specific gravities and boiling-points, being in like manner condensation-products of $2(C_5H_8)$. In all of these, also, the points of fusion and of ebullition help to guide us. When, however, we come to deal with non-volatile solids, like calcite, this mode of determination is no longer open to us, and we cannot say whether in the calcites the unit is $CCaO_3$ or a multiple thereof by two, four, or eight; whether, in short, the formula for the calcite with density 2.688 is to be written $576(CCaO_3)$, $288(C_2Ca_2O_6)$, $144(C_4Ca_4O_{12})$, or $72(C_8Ca_8O_{24})$. Could we observe sharply marked intervals in the density of various calcites, corresponding to certain multiples of $CCaO_3$, they would furnish the data for the determination of the problem. But in this case we may, from the absence of such intervals, infer the probable intervention of the principle of crystalline admixtures among isomorphous species, giving rise to calcites of intermediate densities. This principle, first clearly applied in mineralogy by Von Waltershausen, in 1853, and by the writer, more fully, and independently, in 1854 (**6**), is of importance within the limits then assigned to it; though, as has been elsewhere shown, its application has been misconceived by Tschermak, who some years afterwards adopted it.[1]

§ 67. In the equation $p \div d = v$, it will be noted that the value of $p$ is, for all gaseous or volatile species, deduced directly from the observed density of the gas or vapor, and represents the specific gravity of such species, hydrogen gas being unity ($H_2 = 2.0$); the same value, for non-volatile species, being assumed as corresponding or equivalent thereto. As examples, for the gaseous carbon dioxyd $CO_2$, we find $p = 44$ ($O = 16$); while for the related silicon dioxyd, $SiO_2$, which has only been observed in solid, dense, polymeric forms, such as quartz and tridymite, we assume $p = 60$, which is the equivalent weight of the unknown normal gaseous silicon dioxyd. For calcite, in like manner, we assume the formula $CCaO_3$ with a value for $p = 100$, which is the equiva-

---

[1] See Mineral Physiology and Physiography, pp. 294-296, 342.

lent weight of a gas or vapor less dense than that of amylic alcohol.

On the other hand, $d$ represents the density as computed for liquids and solids; for all of which that of water at 4° is taken as unity. The relation $d:p$ is, then, that of the density of water to that of hydrogen gas,—the two units of specific gravity,—or, in general terms, is that of liquid and solid to that of gas and vapor; while $v$ shows the contraction in passing from the gaseous to the liquid or solid state or, in other words, is the reciprocal of the coefficient of condensation. For water, in which the value of $d=1.000$, it is evident that $d=v$, or, in other words, that the reciprocal is identical with the gaseous equivalent weight. We have selected water as the first example, precisely because its specific gravity at 4° ($=1.000$) is made the unit of density for all liquids and solids; from which it follows that the equivalent weight of its vapor, $p$ (which is its specific gravity, $H=1.000$), and its reciprocal of condensation, $v$ (taking now the corrected equivalent weight of oxygen as being 15.96) are alike 17.96. In the case of ice, however, if we take $d = 0.918$, we have $v = 19.564$. We thus obtain:—

Water = 1192($H_2O$) . . 1192 × 17.960 = 21,408.
Ice = 1094($H_2O$) . . . 1094 × 19.564 = 21,403.

§ 68. The question to be settled in fixing the equivalent of water is to know its specific gravity as compared with that of the same volume of hydrogen at the standard temperature and pressure of 0° and 760 mm., which is accurately known, or with that of water-vapor at the same temperature and pressure. These relations are conveniently shown in the following table:—

| Species. | Formula. | Grams in 1 Litre 0° and 760 mm. | Equivalent Weight. |
|---|---|---|---|
| Hydrogen | $H_2$ | 0.089578 | 2.0000 |
| Steam | $H_2O$ | 0.8044 | 17.9600 |
| Water | 1192($H_2O$) | 999.8700 | 21,408.0000 |

It will be remembered that the contraction of water from 0° to 4° suffices to raise the weight of the litre to 1000 grams.

§ 69. To give a farther illustration: the hydrocarbon vapor, butane, $C_4H_{10} = 58$, condenses below 10° to a liquid having at 0° an observed specific gravity of 0.600, which corresponds to an equivalent weight of 12,840, compared with water as 21,400. This is not far from $222(C_4H_{10}) = 12,876$, which, at its point of ebullition, should have a specific gravity of 0.601. While the reciprocal of the coefficient of condensation for steam, $v = 17.96$, that for the gaseous hydrocarbon butane, with its observed density, is

$$0.600 : 1.000 :: 58 : v = 96.66.$$

The value $v$ in the formula $p \div d = v$, being the reciprocal of the coefficient of the condensation suffered by the gaseous species (or the hypothetical unit), having its combining weight (or, in other words, its specific gravity, hydrogen gas being unity) represented by $p$, in passing to a specific gravity calculated for water at 4° as unity, and represented by $d$. We may, therefore, for brevity, designate the value $v$ thus calculated for the condensed species as its *reciprocal number*.

The density of solids, as of liquid species, should for comparison, as already pointed out in

1853 (§ 21), be taken at the temperature of their formation, or, rather, at the highest temperature which they will sustain without chemical change. This point is, however, difficult to fix for many solids; and, as the coefficient of expansion varies for different species, and is at most not very considerable, we take, for the present purposes of study, the densities of these as already determined at ordinary temperatures.

# CHAPTER XII.

### A HISTORICAL RETROSPECT.

§ 70. OF the fundamental conceptions embodied in the scheme of chemical philosophy set forth in the preceding pages, there are two which have a historic interest apart from the record of them in the writer's earlier papers. These are: (1.) The doctrine of high equivalent weights and complex formulas for all liquid and solid species, dependent on homogeneous integration or so-called polymerization, and the direct connection of this, not only with hardness and insolubility, or chemical indifference, but with specific gravity; which, it has been maintained, varies for liquids and solids, as well as for gases and vapors, directly as the equivalent weight. (2.) The doctrine that liquefaction, solidification, vaporization, and the condensation of vapors, as well as solution, — in short, all changes of state in inorganic species,—are chemical changes. It is proposed in the present chapter to notice how far these conceptions have been entertained, and to what extent they have been advanced, by others previous to or during the period of the development of the scheme which, after a growth of thirty-eight years, is now for the first time resumed and stated in a completed form.

§ 71. The conceptions of high equivalent weights and of polymerism in solid species, as deductions from thermo-chemical studies, were indicated in a general way by Favre and Silbermann, in 1847, whose conclusions were cited by the writer in his more detailed exposition of the view, in 1853 (§ 20). Graham, also, clearly stated the notion of such polymers, though only as existing in solutions, in 1849, as a deduction from his studies in diffusion, as will be shown farther on, in § 83. Their existence was maintained on other grounds, by the writer, for liquid and in solid species, both complex and elemental, in 1848, and again in 1853, 1867, and 1874, as already recounted in preceding chapters.

Spencer Pickering, who does not seem to be aware of their previous history, has set forth similar views, but only so far as regards polymerism, and the consequent high equivalent weights, in an essay on The Molecular Weights of Liquids and Solids, in 1885.[1] Therein, after declaring that "considerations based on the crystalline form and other physical properties of bodies force upon us the conclusion that liquid and solid molecules are in all probability of a very complicated nature, — certainly, more complex than gaseous molecules," — he writes that "the molecule of the chemist is not necessarily identical with the molecule of the physicist"; and supposes that chemical molecules may "agglomerate and act in unison as regards certain physical forces." That such an "agglomerate does not act as a unit towards chemical forces" shows, according to Pickering, that "the force which unites the individuals constituting it is not chemical force, or is chemical force of such a weak nature" as to oppose no

[1] Proceedings of British Association for the Advancement of Science, in the Chemical News for November 13 and 20, 1885.

perceptible resistance to ordinary chemical agents. We have here, in a mechanical form, and in the language of the atomic hypothesis, a reproduction of the views put forward by the present writer, in his early paper of 1853, as to the polymerism and high equivalent weight of liquid and solid species.

§ 72. Pickering's paper was presented to the British Association for the Advancement of Science in 1885, at the same time with one by Guthrie on Physical Molecular Equivalents. He therein calls attention to bodies previously studied by himself, and designated by him as cryohydrates and subcryohydrates; which may be described as very fusible, definite crystalline compounds, containing many portions of $H_2O$ to a single portion of a salt or alkaloid. In these potassium nitrate is said to be combined with $44.6(H_2O)$, and potassium sulphate with $114.2(H_2O)$; while diethylamine forms a well defined and crystalline compound with $27(H_2O)$. Guthrie conceives that in these compounds the "constituents are not in the ratio of any simple multiples of their chemical equivalents." A similar conception is by him extended to certain metallic alloys, which, from their ready fusibility, he designates, like these cryohydrates, as "eutectic."[1]

§ 73. On the other hand, Messrs. Tilden and Shenstone, in their studies of the great solubility in water of salts when near their melting-points, have described compounds which, unlike the cryohydrates, are homogeneous liquids, containing, in the case of somewhat fusible compounds, like silver nitrate and benzoic acid, very small proportions of water; from which the experimenters are led to conjecture for bodies an indefinite solubility in water under proper conditions of temperature and pressure.[2] It is, however, easy to understand from the high equivalent weights of such dense liquids (which from their viscosity are probably colloidal) that compounds with water may exist in which the proportion of the latter, though definite, is so small a fraction that it would be neglected in the ordinary processes of analysis.

[1] Fred. Guthrie, on Physical Molecular Equivalents, Nature, Nov. 6, 1885; also on Eutexia, L., E., and D. Philosophical Magazine, 1884, xvii. 262.

[2] Philosophical Transactions, 1884; part I., pp. 23-36.

The subject of these compounds is discussed by the writer in his Mineral Physiology and Physiography, pp. 221–222, and also p. 245, *note;* where the geological significance of small portions of water in combination under high pressure in molten silicated rocks is insisted upon.

These two cases of apparently homogeneous compounds, — on the one hand, liquids existing at high temperatures, with very small proportions, and, on the other, solids at low temperatures, with very large proportions of water, — are like illustrations of the complex formulas and high equivalent weights which we have so long maintained. So far from being, as conjectured by Pickering, and by Guthrie, compounds in some sense outside of the domain of chemistry, their constitution, as was said of similar cases in 1867, does not present "a deviation from the law of definite proportions," but "is only an expression of that law in a higher form." (**10.**)

§ 74. A broader view, and a more complete comprehension of the great problem under consideration, is shown in a remarkable paper by Professor Louis Henry, of the Catholic University of Louvain, published in August, 1885, on The Polymerization of Metallic Oxyds,[1] in which, though without any reference to the present writer's publications, will be found developed alike the conception of polymerism in liquid and solid species, and the dependence thereon, not only of hardness and chemical indifference, but of specific gravity. In fact, most of the important conclusions announced in the author's publications from 1848 to 1867, inclusive, are set forth in the language of the atomic hypothesis by Henry in his recent paper.

He explains the generation of complex oxyds by the union of several equivalents of hydrates, and the successive elimination of equivalents of water, as taught by Adolphe Wurtz in 1860, in discussing the origin of polysilicates.[2]

[1] London, Edinburgh, and Dublin Philosophical Magazine (5) xx. 81–117. This essay escaped the notice of the writer until October, 1886; otherwise it would have been noticed in referring in Science for Sept. 10, 1886, to the paper of Spencer Pickering, when the mode of fixing the value of the coefficient of condensation for liquids and solids was first set forth, as shown in § 46.

[2] See the author's Mineral Physiology and Physiography, p. 298.

While the chlorids of carbon, silicon, titanium, and aluminium are species readily converted into vapors of normal density; and while carbonic dioxyd is a gas, only changed into a liquid form under great pressure, the oxyds of silicon, titanium, and aluminium are fixed and solid bodies, of greater or less density and hardness, and are polymers of the elemental oxyds. These, in their more highly condensed forms, "show a greater resistance to the action of chemical agents than in their former condition, in some cases even entirely resisting the action of acids and alkalies." Henry concludes that "the true oxyds, which are really comparable with the chlorids, are unknown, and that we possess only polymers of these, $(ROx)n$." The condensation of these varies, "but appears to attain its maximum in certain fixed and very infusible oxyds, such as silica, alumina, etc. . . . This enormous condensation of their molecules may possibly be the cause of the greater resistance which they offer to the action of simple chemical agents, such as hydrogen, carbon, sulphur, etc."

§ 75. Henry next suggests that "the specific

gravity of a solid body doubtless depends in a great measure on the state of aggregation, and is also intimately connected with the composition of the body, and the size of the molecular weight"; adding "the density then is, to a certain degree, a function of the molecular weight." Having reached this point, he proceeds as follows: "What, now, is the true value of the coefficient $n$ of polymerization? What is the real molecular formula of these polymeric oxyds? These questions are, doubtless, of great interest; but it should be stated at once that it is absolutely impossible to give a direct answer. I do not know of any fact which would allow us to assign an absolute value to the coefficient $n$ of polymerization. . . . So far as facts will permit a conclusion, we may affirm that in most cases this number is very high, although different for different oxyds."

§ 76. The problem of the value of the condensation in passing from the gaseous to the solid state, as presented by Henry, in 1885, is thus precisely that put forth by the present writer in 1853, when he asked for a determination of the coefficient of condensation, and at the same time affirmed more positively than is done by his successor in 1885, that the density in liquids and solids varies directly as the equivalent or so-called molecular weight. (§ 41.) The direct relations of this condensation to hardness, and to chemical indifference, were, moreover, as we have seen in preceding chapters, clearly formulated by the present writer, in 1863, and even in 1848 were suggested in the case of the forms of phosphorus, etc. The problem of fixing the coefficient of condensation thus put forth, although not answered, in 1853, and again propounded, and declared unsolved, by Henry, in 1885,[1] has, it is believed, been successfully resolved in the preceding chapters.

[1] Henry has given, in his paper, the following valuable classification of oxyds, considered with reference to polymerization: —
  A. Normal oxyds, as $SO_2$, $CO_2$, NO.
  B. Polymerized oxyds, consisting of *n* molecules of the normal oxyds united.
    1. Volatile oxyds, totally depolymerized by heating, and converted into the normal oxyd.
      *a.* Completely depolymerized on volatilization, and yielding a vapor of normal density, as $SO_3$, $OsO_4$, methylene oxyd $(CH_2O)n$, metaldehyde$(C_2H_4O)_3$.

§ 77. We must here notice an essay which appeared in 1876, entitled Geometrical Chemistry, by Prof. Henry Wurtz,[1] of New York, in which he insists on the great significance of density, and of the variations therein, as observed in liquid and solid species. He asserts (evidently excepting the temporary changes which depend on variations in temperature)

> b. Incompletely depolymerized at the moment of volatilization, the process being completed progressively, as the temperature rises. The vapor-density of these gradually diminishes with increase of temperature, up to a certain point, beyond which it becomes constant, and corresponds to the normal oxyds; $N_2O_4$, fatty acids, and paraldehyde, which last, at low temperatures, corresponds to $(C_2H_4O)_3$, but at higher temperatures becomes $(C_2H_4O)$. [In this category, we may place the elemental species iodine, bromine, and chlorine.]
> 
> 2. Volatile oxyds, incompletely depolymerized at temperatures to which they have been subjected, as $(As_2O_3)_2$. The volatilization of solid arsenious oxyd, at 200°, without fusion, is the change of $(As_2O_3)n$ into $(As_2O_3)_2$.
> 3. Oxyds not capable of depolymerization.
>    a. Fixed oxyds, like $SiO_2$, $MgO$, $Al_2O_3$, $Fe_2O_3$.
>    b. Oxyds volatile or decomposed by heat, as $HgO$, $Ag_2O$, $MnO_2$, $CrO_3$; also, organic oxyds decomposed by heating.

[1] Geometrical Chemistry, by Henry Wurtz, pp. 73, reprinted from the American Chemist for March, 1876.

that "change of volume or of density in a specific liquid, or a specifically solid body, whether elementary or compound, — even when occurring without change of crystalline system, — is an infallible indication of change of molecular structure, and, consequently, of chemical nature." Without attempting to grapple with the problem of the direct relation of density in such bodies to their equivalent weights, as propounded by the writer in 1853, and again by Louis Henry in 1885, Wurtz, nevertheless, sought to frame, in accordance with the atomic hypothesis, an explanation of the densities of liquid and solid species. To this end he constructed for these chemical formulas, which, from their frequent complexity, seem to imply polymerism and high equivalent weights. We find, however, on farther examination, that the weights adopted by him do not represent those deducible from these formulas, but are the simplest combining numbers, — hydrogen being unity, — multiplied, however, by 1000, to avoid fractions.

§ 78. Regarding chemical species as built up by the agglomeration or juxtaposition of elemental molecules, Wurtz supposes that similar molecules entering into a compound may possess unlike volumes, and to this attributes the variations in the density of isomeric species. The elemental molecules in his hypothesis have different diameters, represented by integers from 12 to 40, the cubes of these numbers being the volumes of the molecules. These diameters for the metals, hydrogen, and phosphorus, are very variable; while for sulphur, chlorine, bromine, and iodine, they are pretty constantly 28 or 24; the latter being the diameter for chlorine in chlorates and in all monad or dyad chlorids. This being conceded, he applies to any liquid or solid species the formula $p \div d = v$, and thus gets what has been called its molecular volume, multiplied, like its combining weight $p$, by 1000. Taking now as an example, liquid chlorhydric acid, with a combining weight for $HCl = p = 36,500$, and an observed density of 1.27, we get $v = 28,516$. Deducting from this the volume of $Cl = 24^3 = 13,824$, there remains for the volume of $H = 14,692$, which, as he notes, is "almost identical with the mean figure between $24^3$ and $25^3$," whence he concludes that "liquid hydrochloric acid must be $H_2Cl_2$, containing H with the two diameters 24 and 25," and that its "volumic or molecular formula" must be represented accordingly. But the volume, as deduced from this formula $= (24^3 \times 3) + (25^3) = 57,097$, is, as near as the method will admit, double the value of $v$; so that the "molecular formula," as understood by Wurtz, corresponds not to $H_2Cl_2$, with two diameters for the two molecules of $H_2$, but to a mean between HCl with $H = 24$ and HCl with $H = 25$.

§ 79. In like manner, sodium chlorid with density 2.25 is represented by NaCl, in which the diameter of the molecule of sodium $= 23$; while that with density 2.18 is supposed to be represented by $Na_2Cl_2$, in which we have sodium with the two diameters of 23 and 24. The same chlorid with density 2.05 is $Na_2Cl_2$, having for sodium the two diameters 24 and 25; potassium chlorid, with density 2.01, being $K_2Cl_2$, with the diameters 28 and 29 for the two potassium molecules. Similarly, phosphorus with density 1.797 is supposed to consist of one molecule with a diameter of 25, and five molecules with a diameter of 26; while metalloidal phosphorus, with density 2.297, includes one molecule of 23 and four of 24. In these cases of apparently complex formulas, however, as in the case of chlorhydric acid, the value adopted for $p$ is not a multiple, but the number representing NaCl or P. Thus, for metalloidal phosphorus, described above as made up of five molecules of P, and with a density of 2.297, we have, with $p = 31,000$, a value for $v = 13,495$. But $(24^3) \times 4 + (23^3) = 67,463 \div 5 = v = 13,492$; while $p \div v = d = 2.297$, the observed density.

§ 80. By thus adding together at pleasure the cubes of any number of integers between 12 and 40, and dividing the sum by this number, in order to obtain a mean value wherewith to divide the combining weight of the species, it is evident that we may arrive at a near approximation to the value of $v$ as found for any observed density of a given species, and therefrom get by calculation a closely corresponding density. The assumption, however, that $v$ represents the volume of the composite molecule generated by the putting together of elemental molecules is, as we have sought to show, unfounded; while the other assumption that the diameters of these are represented by the integers above indicated is gratuitous. There is no apparent relation in the nature of things between these integers and the amount of contraction suffered, for example, by the gaseous species HCl, with a specific gravity $=$ 36.5 (hydrogen being unity), in passing into the liquid species, having a specific gravity of 1.27 (water being unity); and we must regard the calculations of density, and of chemical constitution, arrived at by the above method as illusory and essentially fallacious.

Moreover, as we have elsewhere said, while calling attention to the importance of the study of the densities of liquids and solids, so well insisted upon by Wurtz, "the conception that the chemical elements enter as such into combination, and there retain their volumes, appears to be inadmissible in chemical philosophy." (**16**.)

§ 81. Proceeding now to consider the history of the conception that all changes of state, such as solution, fusion, solidification, vaporization, and the condensation of vapors, are chemical in their nature, we are led to review the phenomena of gaseous and liquid diffusion, as definitely set forth by Graham, in 1849. The diffusion of salts from their solutions into pure water was by him compared to the diffusion of gases, and to the evaporation of liquids into the atmosphere. "The analogy of liquid diffusion to gaseous diffusion is borne out in every character of the former which has been examined," according to Graham; while diffusion was farther said to be "a property of a fundamental character, upon which other properties depend, like the volatility of substances." "The number of species that are soluble, and therefore diffusible, appears to be much greater than the number of volatile bodies." "Separations, both mechanical and chemical (decompositions), are produced by liquid, as well as by gaseous diffusion." "Liquid diffusion, as well as gaseous evaporation, may produce chemical decompositions." "The diffusion of a salt appears to try its tendencies to decomposition very severely."

§ 82. The above citations are from Graham's

Bakerian Lecture on Diffusion, in 1849. From his extended studies of this subject up to that date, he was led to conclude that there exist groups of equi-diffusive substances, which coincide, in many cases, with the isomorphous groups, but, on the whole, are more comprehensive than the latter. Moreover, for several groups of salts, it was found that the squares of the times of equal diffusion from solutions of the same strength stand to each other in a simple numerical relation. The squares of the times of equal diffusion of gases, as Graham had shown, are to one another in the ratio of their densities. From this, by analogy, he inferred that the molecules of these several salts, as they exist in solution, possess densities which are to one another as the squares of their times of diffusion. These he called their *solution-densities*, which, for the sulphate, nitrate, and hydrate of potassium, for example, were found to be as $4 : 2 : 1$; being the squares of 2, 1.4142 and 1, the numbers representing the relative times occupied in the diffusion of equal weights of these three species.

§ 83. Graham thus constructed a scale of

solution-densities which, in his words, "are suggested by the different diffusibilities of salts, and to which alone, guided by the analogies of gaseous diffusion, we can refer these diffusibilities. Liquid diffusion thus supplies the densities for a new kind of molecules, but [supplies] nothing more about them." "The fact that the relations in diffusion of different substances refer to equal weights of these substances, and not to their atomic weights or equivalents, is one that reaches to the very bases of molecular chemistry. The relation most frequently perceived is that of equality. In liquid diffusion we appear to deal no longer with chemical equivalents, or the Daltonian atoms, but with masses even more simply related than these in weight. Founding still upon the chemical atoms, we may suppose that they can group themselves together in such numbers as to form new and larger molecules, of equal weight for different substances; or, if not of equal weight, of weights which appear to have a simple relation to each other. It is this new class of molecules which appears to play a part in solubility, and in liquid diffu-

# A Historical Retrospect. 131

sion, and not the atoms of chemical combination."

The reader can judge how far this generalization of Graham, as to an apparent polymerization of species in solution, deduced from their relative rates of aqueous diffusion, led the writer naturally to the broader doctrine of polymerization alike in liquid and solid species, as put forth by him in 1853, and again in 1867. Here, also, is the view, reiterated by Spencer Pickering, of the formation of physical molecules by the union of many chemical molecules.

§ 84. Continuing his studies in diffusion, Graham wrote farther, in 1861: "So similar in character to volatility is the diffusive power possessed by all liquid substances, that we may fairly reckon upon a class of analogous analytical resources to arise from it," — a conception fully realized in his admirable researches, published at the time, on Liquid Diffusion applied to Analysis. He added: "The range in the degree of diffusible mobility exhibited by different substances appears to be as wide as the scale of vapor-densities." Of potassium hydrate it was then said that it has double the velocity of diffusion of potassium sulphate, and this double that of sugar, alcohol, and magnesium sulphate; all of which, however, as regards diffusion, he compared to the more volatile bodies — the comparatively fixed, or non-volatile bodies, being represented by the colloids. Of these he remarks that, "while the rigidity of crystalline structure shuts out external impressions, the softness of the gelatinous colloid partakes of fluidity, and enables the colloid to become a medium for liquid diffusion, like water itself." "Colloids are characterized by mutability." "Their existence is a continual metastasis."[1] His conjecture that colloids have perhaps a higher equivalent weight than crystalloids, has apparently no substantial basis. We have elsewhere (§ 30) noticed the earlier observations of Breithaupt on colloids, both of aqueous and igneous origin, called by him porodic bodies. The relation of such substances to the process of diffusion was, however, unknown to Breithaupt, and was discovered by Graham.

---

[1] The papers of Graham above cited, published in the Philosophical Transactions for 1850 and 1861, are reprinted in his Chemical and Physical Researches, 1876, pp. 444–600.

§ 85. As early as 1847, in a paper already noticed (§ 20), Favre and Silbermann,[1] had described the loss and fixation of water in the efflorescence and deliquescence caused by the action of the atmosphere on certain bodies, and the precipitation by heat of anhydrous sodium sulphate from its aqueous solution, as alike chemical processes. As facts in the same order, Graham had shown, in 1849, that both liquid and gaseous diffusion may produce chemical decomposition, and had compared these processes to volatilization. It was reserved for Henri Sainte-Claire Deville to take another step in this new way by his studies in chemical dissociation, the publication of which was begun in 1857, when he described the decomposition by heat of potassium hydrate, and showed the partial decomposition of water at temperatures much lower than had hitherto been suspected.[2] In 1860, he said of dissociations by heat, that we are "authorized in assimilating these phenomena to those of the ebulli-

---

[1] Comptes Rendus de l'Acad. des Sciences, 1847, xxiv. 1081–1090.

[2] *Ibid.*, 1857; xlv. 857.

tion of liquids, or the vaporization of non-liquefiable solids."[1] The action of metallic points in promoting ebullition was then by him compáred to that of manganese oxyd in fused potassium chlorate.

§ 86. Returning to the subject in 1863,[2] in farther discussion of the decomposition of water by heat, Deville wrote: "The comparisons between the effects of *cohesion* and of *affinity*, which are so instructive for liquid and solid bodies, are maintained in the inverse phenomena of *volatilization* and *decomposition*. Admitting this comparison, we see that the phenomena of the decomposition of bodies at a relatively low temperature, or dissociation, corresponds to the vaporization of a liquid at a temperature below its boiling-point; and that the quantity of the body dissociated at a given temperature will be proportional to its tension of dissociation expressed in millimetres of mercury; as the quantity of vapor formed above a liquid at a given temperature is proportional to the maximum tension of

[1] L., E., and D. Philos. Mag., 1860; xx. 457.
[2] Comptes Rendus de l'Acad., 1863; lvi. 195–201.

its vapor." "The phenomenon of the decomposition of bodies is in all respects similar to that of the ebullition of liquids, the principal character of which is the invariability of their temperature, whatever the intensity of the source of heat, provided the pressure is constant."

Thus, while looking upon the cohesion of liquids, and their volatilization, as phenomena in some way distinct from those of chemical combination and decomposition, Deville was forced to conclude that they are governed by the same laws; involving a change of state and of thermic relations, and subordinated to the same conditions of temperature and of pressure.

§ 87. In considering, in 1863, the phenomena of dissociation, as then lately studied by Pébal for ammonium chlorid, and by Wanklyn and Robinson for sulphuric acid and phosphorus chlorid, Deville recalls the phenomena of diffusion, as studied by Graham in aqueous solutions, and adds: We have here "a veritable force which leads to the separation of elements, and which must not be neglected in the explanation of the phenomena described by Messrs. Pébal, Wanklyn, and Robinson, since the same reasoning applies to aqueous diffusion and to diffusion in gases or vapors which have different diffusive powers or rates of transpiration. The new agent of decomposition introduced by Graham is so energetic that we can no longer designate as spontaneous the decompositions produced under its influence."[1]

Still later, in 1865,[2] Deville wrote of chemical combination : "We have no knowledge of what combination is; we do not even know what distinguishes it from solution; but we can characterize it as a change of state, marked by new physical or chemical properties, serving to distinguish combination from simple mixture. This change of state is most generally accompanied by a disengagement of latent heat, which approximates combination to the condensation of vapors; but it may also be accompanied by an absorption of latent heat, or cooling, as for explosive bodies like NO and $NCl_3$, which are formed by combination in the nascent state, and give off heat in decomposing; so that the

---

[1] Comptes Rendus de l'Acad., April 20, 1863; lvi. pp. 729-733.

[2] Bulletin de la Société Chimique; iii. 15.

disengagement of heat, the production of cold, or the absence of thermometric effect, proves nothing for or against the fact of combination."

§ 88. Thus we have seen that Favre and Silbermann, in 1847, regarded the separation of salts from solutions by changes of temperature, the fixation of aqueous vapor from the atmosphere by certain bodies, and its loss by others, as all alike chemical changes. Graham, in 1849, insisted upon the likeness of the volatilization of liquids to gaseous and liquid diffusion, and asserted that all of these processes may produce chemical decompositions. Henri Sainte-Claire Deville, in 1857 – 1863, while adopting Graham's conclusions, farther declared that the decomposition of bodies by heat is in all respects similar to that of the evaporation of liquids, and is subordinated to the same laws.

§ 89. The writer, who enjoyed the great advantage of personal intercourse alike with Graham and with Sainte-Claire Deville, and conversed with both of them on the subjects here discussed, had with the latter, whom he had then long known, frequent conferences during the spring and summer of 1867, at his laboratory, at the École Normale, in Paris, when the relations of the homogeneous volatilization of liquids and solids were compared with the phenomena of dissociation. Graham, whom I had known since 1856, was there for a few days our companion; and to both of these I maintained, in accordance with the views in my paper, then recently published (10), that the liquid and solid bodies, water and ice, are polymeric species, distinct from aqueous vapor, their conversion into which is a process of depolymerization, or homogeneous disintegration; a conclusion admitted by both Graham and Deville to be a logical deduction from their own premises, as has already been shown. The views which I have advocated in various publications since 1853, and in the preceding chapters, with regard to the constitution of liquids and solids, their high equivalent weights, and all their relations to gases and vapors, appear to be in fact but the legitimate and necessary consequences of those first put forward by Favre and Silbermann, and continued by Graham, Deville, and myself.

# CHAPTER XIII.

### CONCLUSIONS.

§ 90. THE views maintained in the preceding chapters as a basis for a philosophy of chemistry, may be resumed as follows : —

The chemical process may be defined as the integration or the disintegration of chemical species, resulting in the production of new species, also chemical. Both of these changes may be either homogeneous or heterogeneous; that is to say, the species concerned therein may be elementary, or, if complex, may be like or unlike in centesimal composition.

The generation, by heterogeneous change, of species dissimilar to the parents we have designated metagenesis. In heterogeneous integration two or more unlike species combine to form a new one, and in heterogeneous disintegration a species separates into two or more unlike to each other. The union of hydrogen and carbon to form acetylene, of oxygen and carbon to form carbon dioxyd, the absorption of the latter by sodium hydrate, or of atmospheric water by calcium chlorid, are examples of heterogeneous integration. The loss of water in the evaporation of a saline solution, in the efflorescence of hydrated sodium carbonate, and the disengagement of carbon dioxyd by heat from calcium carbonate, are, in like manner, examples of heterogeneous disintegration. All solution is heterogeneous integration; while the separation of crystals from a solution is heterogeneous disintegration.

What is called double decomposition consists in heterogeneous integration, followed immediately by a disintegration generating two (or more) new species. In other words, two species disappear, and two others take their places; the double decomposition being the result of two consecutive actions, and the heterogeneous disintegration following so closely on integration that it is difficult or impossible to arrest the change at the end of the first stage, so as to isolate the unstable species before it is destroyed. When these processes are confined within narrow limits of time and temperature, the hitherto unexplained operation has been called catalysis, or action by presence.

§ 91. The generation, by homogeneous change, of species like the parents in centesimal composition we have designated metamorphosis. In homogeneous integration the identity of a species is lost in that of a new one of higher equivalent weight. In homogeneous disintegration this process is reversed, and a new species is formed, having the same centesimal composition but a lower equivalent weight. These two cases of metamorphosis correspond to what are called polymerization and depolymerization.

In many instances, as in the vapors of sulphur at different temperatures, and in pentene, and the various aldehydes and their polymers, the ratios between the equivalent weights of the species concerned are very simple; but in other cases, as in the polymers of many elemental species, such as carbon, tin, and phosphorus, in silicon dioxyd, and in native silicates and carbonates, these ratios are more complicated, and do not correspond to very simple fractions or multiples. Hence we infer, in the process of metamorphosis, a temporary homogeneous disintegration of the species into more elemental forms, which, by redintegration, may give rise to new species, either of higher or of lower equivalent weights than the parent.

§ 92. Some species with difficulty assume or maintain a polymeric state, — as, for example, carbon dioxyd, which is liquid only at very low temperatures, or under great pressure, and, moreover, forms a very volatile and unstable solid; while other species, like carbon itself, silicon, and silicon dioxyd, are known to us only in dense polymeric forms, of great though different degrees of condensation. Between these two extremes are bodies which, like phosphorus, pentine, and the aldehydes, yield various polymers of different degrees of complexity and of stability.

All known liquid and solid species, whether elementary or not, are polymers of some simpler species, which, in very many cases, is capable of assuming the form of a gas or vapor. Such vapors themselves are sometimes polymers which, at higher temperatures, may be resolved into still simpler forms by homogeneous differentiation, as in the case of metaldehyde, of sulphur-vapor below 500°, and of iodine-vapor below 800°. Theoretically, we may suppose that even oxygen, nitrogen, and hydrogen may also undergo such resolutions at higher temperatures and under diminished pressure. There are, however, many solid species which, from the facility with which they undergo heterogeneous differentiation, or from their high equivalent weights, cannot exist in the vaporous state.

By integration, alike in metagenesis and metamorphosis, we rise to higher chemical species, and by disintegration descend to simpler ones. If, as we may suppose, all chemical species have come from one primal matter, it follows that the final disintegration of all so-called elementary species must be a homogeneous one, giving rise to that primal substance, which can only be conjectured to be a species much more attenuated than hydrogen, and possibly identical with the matter giving the green line 1474 of the corona, or some other unknown solar element.

§ 93. The condensation of gases into solids

or liquids, and the fusion and volatilization of these, when not attended with heterogeneous disintegration, are examples of metamorphosis. All such changes of state — the conversion of water into steam, the condensation of this again into water, its conversion into ice, and the fusion and evaporation of the latter — are, consequently, chemical processes. The precise nature of the liquid and of the colloid or porodic state, as compared with crystalline individualized solids, is not yet clear. The passage from the colloid to the crystalline, as well as that from the liquid to the crystalline, is generally attended with condensation; while the fusion of a solid is a metamorphosis by expansion. To this, however, the conduct of water, and some other bodies in solidification, offers an exception. The same vaporous species, steam, gives rise by polymerization to the two distinct species, water and ice, having different equivalent weights. Analogies to this are seen in the different forms of liquid and solid phosphorus, and of tin, in the various liquid and solid species of di-pentines or turpentine-oils with different specific gravities and boiling-points, and in a great number of native and artificial oxyds, carbonates, and silicates, as has been shown at some length in Chapter VII. A study of these relations must form the basis of a natural system of mineralogy.

By polymerization, or homogeneous integration, the normal gaseous or volatile species are thus changed into liquids and solids of greater or less degrees of condensation. Of these the denser, other things being equal, are the harder and the more stable or indifferent to chemical changes such as fusion and solution. This relation has been illustrated at length by showing that the hardness and fixity of native sulphids, oxyds, carbonates, and silicates, and their resistance to the action of solvents, augment with the condensation of the species.

§ 94. All chemical changes are subordinated to measure and weight, as shown in the law of definite proportions alike by volume and by weight. The law of numbers is made evident, not only in multiple proportions, and in metamorphosis, but in homologous or progressive series, which, so far from being confined to certain hydrocarbonaceous species, are found throughout all classes of chemical compounds.

In the combination or integration of gases and vapors the volumes uniting are lost in that of the product, there being an identification of volume. The converse is true in decomposition, or disintegration, which is differentiation. The fact that this law of volumes is universal, and applies not only to gases and vapors but to liquids and solids, has been obscured by the false assumption that the volume for solids is a variable and arbitrary quantity, conditioned by crystalline form; so that no constant comparison was possible between the volume of solids and that of gases. Moreover, there have hitherto been two distinct units of comparison, the one for gases, and the other for liquids and solids. The weight of a given volume of a gas or vapor as compared with that of the same volume of hydrogen gas at standard temperature and pressure, — which is its equivalent weight, — is, at the same time, its specific gravity, hydrogen being unity. The weight of a similar volume of a liquid or solid species is, on the contrary, compared with that of water;

and its equivalent weight also varies as its specific gravity, the unit of which, $1192(H_2O) = 21,408$ (§§ 46, 61, and, farther on, § 137).

§ 95. In order to compare liquids and solids with gases and vapors, it is necessary to refer them to a common standard of density, or, in other words, to determine the relation of density between water and hydrogen gas. This is the relation of $d:p$ in the proportion $d:p :: 1 : v$, provided $d$ be the density of water at 4°, and $p$ that of hydrogen at 0°. To fix this relation, we compare the weight of a given volume (one litre) of hydrogen gas at 0° and 760 mm. with that of the same volume of water at the temperature of 100°, remembering, meanwhile, that this liquid at 4° has a weight of 1000 grams to the litre. We have thus the means of determining the specific gravity of hydrogen gas at 0°, and that of water at 100°, — the temperature of its ebullition and its condensation at 760 mm., — as compared with water at its maximum density, which has been assumed as the unit of specific gravity for liquids and solids.

§ 96. With gases and vapors, the equivalent

weights of which are directly as their densities, it is evident that in the proportion $d : p : : 1 : v$, we have $d = p$. But when this formula is applied to liquids and solids, while $p$ still represents alike the equivalent weight and the density (hydrogen being unity) of the normal gaseous species, $d$ is made to represent the density of some polymer thereof. In the case of water itself, where $p = 17.96$ is both the equivalent weight and the density of steam, $H_2O$ (hydrogen gas, $H_2 = 2.0000$), $d$ represents the weight of that liquid at its maximum density (which is assumed as the unit of specific gravity). We have then :—

$$1.0000 : 17.96 :: 1 : v = 17.96.$$

The number thus found is the reciprocal of the coefficient of the condensation suffered by $p$ in passing from the gaseous condition to the density in question. To determine the amount of this condensation we have simply to compare the weight of equal volumes of the gaseous $p$ at standard temperature and pressure, and of the liquid $d$ at the temperature of its condensation under the same pressure; in other words, knowing the weight of the litre of water-vapor at 0° and 760 mm., to compare it with the weight of a litre of water at 100°. In this way we have found (§ 46) that water is formed by the condensation into one volume of 1192 volumes of $H_2O$, at standard temperature and pressure.

In the proportion $d:p::1:v$, as applied to any liquid or solid species, $p$ does not represent its real equivalent weight, but the weight of the normal species which, by its condensation, gives rise to the liquid or solid. The value of $v$ is thus that of the reciprocal of the coefficient of condensation. The equivalent weight of water (and of any other species having a specific gravity of 1.000) being as given above, we have $21,408 \div v =$ the coefficient of condensation.

From these data, remembering that the law of volumes is universal, and that equivalent weights are but the weights of equal volumes, it is clear that the equivalent weight varies directly as the specific gravity; so that, this being known, it is easy to calculate the true equivalent weight of any species.

§ 97. The phenomena of temperature, radiant energy, and electricity, apparent in chemical changes, do not belong to chemism, but to dynamics. To the dynamical history of species pertain all questions as to the atomic or molecular constitution of matter, on the one hand, or its infinite divisibility, on the other. Continued subdivisions cannot effect the destruction of a chemical species, which is an integer or unit, in which the existence, as such, of none of the species which may be obtained by its disintegration can be affirmed. No hypothesis as to an atomic or molecular structure, framed to explain dynamic phenomena in any given species, can, therefore, be legitimately extended to explain the generation of those species which may be derived from it either by integration or disintegration. For these reasons it is conceived that all such hypotheses, however useful they may be found in the explanation of certain dynamical phenomena, are foreign to the domain of chemistry, and should hold no place in the theory of the science.

§ 98. The chemical process is subordinated to the influences of pressure, temperature, and radiant energy. Pressure influences chemical change, as seen in its relations to vaporization, and to heterogeneous dissociation, as well as to solution and to fusion; all of which it aids or restrains according as these are attended by condensation or expansion. A remarkable case is seen in its effect on the brittle species of tin, which, by pressure, undergoes a ready metamorphosis into the denser species. In like manner, as W. Spring has shown, amorphous and prismatic sulphur pass, under pressure, into the denser and more fusible prismatic species. Heterogeneous integration is moreover directly effected thereby, as in Spring's experiments upon mixtures of metals with sulphur, or with each other. (See farther §§ 126–130.)

Heat promotes union within certain limits, as seen in the solution of many substances in water, and in combustion; where, however, it may act by first causing dissociation, since nearly all species known are disintegrated at sufficiently high temperatures, when the tendency to integration is nullified, and matter becomes indifferent to chemical change. Heat is the universal disintegrator; and change of state comes in the primal matter by reduction of temperature, and increase of pressure, with which chemical integration begins. Radiant energy also effects chemical change, as seen in the action of light; and the electric current causes chemical disintegration, in a way not yet well understood.

Chemism, however, is not to be confounded with any of the dynamic agencies just mentioned. It is one of the manifestations of an energy pertaining to matter, which inclines it to integration or disintegration, as conditions favor the one or the other mode of change. The energy thus displayed in change of state appears also in these dynamic activities, and is one and the same, whether manifested on the plane of dynamics, on that of chemism, or on the higher plane of biotics; but it is an error to confound either dynamical or biotical with chemical activity.

# CHAPTER XIV.

### SUPPLEMENT.

§ 99. THE design of the present chapter is two-fold : first, to supply certain omissions in the historical sketch of the development of the views set forth in the preceding, chapters ; and, second, to give to these views a farther extension, and thus realize, to some extent, the idea — suggested in the preface to the first edition of this volume — of building, upon the basis which we have sought to establish, a new chemistry. Two papers, with this double object, have since been published, which, numbered in sequence to those enumerated in § 1, are as follows : —

**18.** CHEMICAL INTEGRATION ; read before the National Academy of Sciences, at Washington, April 19, 1887, and published in the American Journal of Science for August and in the Chemical News for September 23 and 30 of the same year.

**19.** INTEGRAL WEIGHTS IN CHEMISTRY; read before section B of the British Association for the Advancement of Science, at Manchester, September 1, 1887, and published in the London, Edinburgh, and Dublin Philosophical Magazine for October, 1887.

To these must be added two other papers unpublished, entitled, CHEMISM AS RELATED TO TEMPERATURE AND PRESSURE, and PROGRESSIVE SERIES IN CHEMISTRY; both read before the National Academy of Sciences, at New York, November 10, 1887. It is proposed to give, in the following pages, a synopsis of the views embodied in the four papers just named, the extracts from the published ones being designated by affixing to them the corresponding number.

§ 100. The views enunciated in 1853, in the paper numbered **4,** which have been so frequently cited in the preceding pages, were resumed, in part, in a note read before the Académie des Sciences, in Paris, in 1855, SUR LES VOLUMES ATOMIQUES, wherein, "after saying that, 'since the chemical combination of two bodies is to be regarded as an interpenetration of masses, and not as a juxtaposition of molecules, the atomic hypothesis is not necessary for the explanation of the law of equivalent weights,' it was added : 'Their densities thus furnish us a means of fixing the equivalent weights of gases, and of bodies which volatilize without decomposition, and it remains to be determined if some law as simple as that of Gay-Lussac will not permit us to fix, by similar means, the equivalents of solid and non-volatile bodies.'[1]" (18.)

In the same note, the question of progressive and homologous series was farther considered, and it was said : "'These homologous relations, far from being limited to carbon compounds, are but examples of that numeric harmony seen

[1] "Puisque la combinaison chimique de deux corps doit être regardée comme une interpénétration des masses et non pas une juxtaposition des molécules, l'hypothèse atomique n'est pas nécessaire pour expliquer la loi des poids équivalents."
... "Les densités nous fournissent ainsi un moyen de fixer les poids équivalents des gas et des corps qui sont vaporizables sans décomposition, et il restait à déterminer si quelque loi aussi simple que celle de Gay-Lussac ne nous permettrait pas de fixer par un moyen semblable les équivalents des corps solides et non volatiles." Comptes Rendus de l'Académie des Sciences, xli. 77–81.

by Laurent, and recognized by Dumas in the equivalents of the elements, which will become for chemistry a principle as wide in its application as those of atomic weights and volumes.' Of the writer's studies of the latter, it was further said by him that they were undertaken 'in the hope of giving to mineral chemistry something of that exactitude which organic chemistry already possesses.' "[1] **(18.)**

§ 101. In a later note, presented to the same Academy, also in 1855, SUR LES RAPPORTS ENTRE QUELQUES COMPOSÉS DIFFÉRENTS PAR $H_2$ ET $O_2$, "after recalling the above conclusions as to homologous series, it was maintained that similar relations may exist between bodies differing in their proportions of oxygen or of hydrogen. In support of this thesis, as regards oxygen, were then compared, alike chemically

---

[1] "Ces rapports d'homologie, loin d'être limités aux composés de carbone, ne sont que des exemples de cette harmonie numérique que voyait déjà Laurent, que Dumas a reconnue dans les équivalents des éléments, et qui deviendra pour la chimie un principe d'une application aussi large que ceux des poids et des volumes atomiques." . . . "Dans l'espoir de donner à la chimie minérale quelque chose de cette exactitude que possède déjà la chimie organique." *Ibid.*

and crystallographically, malic and tartaric acids, chlorates and perchlorates, sulphates, carbonates and sulphato-carbonates. As regards hydrogen, the compound ammonias, $NH_3 + n(CH_2)$, were compared with the analogous base, piperidine, belonging to a series $NH + n(CH_2)$, and with arsine or methyl arsenid, $AsH + CH_2$. These facts, and other considerations, it was then argued, 'lead us to admit an intimate relation between bodies differing by $H_2$,' as well as by $O_2$ $(O = 8)$.[1]" (**18.**)

§ 102. "A farther illustration of this extension of the conception of progressive series was soon afterwards afforded by the studies of J. P. Cooke on the crystallized alloys of zinc and antimony, which, with similar crystalline forms, present considerable variations in the proportions of their constituents, leading him to the conclusion that 'zinc and antimony are capable of uniting in other proportions than those of their chemical equivalents; or, in other words, that the law of definite proportions is not so

---

[1] "Nous portent à admettre un rapport intime entre les corps différents par $H_2$." Comptes Rendus de l'Académie des Sciences, xli. 1167.

absolute as has been hitherto supposed.'[1] In commenting on these results, as they had been set forth by Cooke in 1860, it was said by the writer in 1874 : 'These alloys of varying composition are to be regarded, in part, as examples of a progressive series of isomorphous compounds of antimony and zinc, of high equivalent, differing from each other by $n\text{Zn}_2$, and in part, doubtless, as crystalline mixtures of these isomorphous homologous species. The principle embodied in the conception advanced by Professor Cooke, and rightly regarded by him as of great importance to a correct science of mineralogy, he has named allomerism. It is evidently a case of homologous and isomorphous relations between members of a progressive series — a general principle upon which I have insisted throughout the pages of this paper [one in the Compte Rendu of the French Academy of Sciences for June 29, 1863, there reprinted], and which includes the polymeric isomorphism of Scheerer.'"[2] **(18.)**

---

[1] Cooke, American Journal of Science, 1883, xxvi. 310–316, resuming the conclusions of his studies of 1855 and 1860.

[2] Hunt, Chemical and Geological Essays, 447.

§ 103. These observations of Cooke have already been briefly noticed in § 24, where, also, allusion has been made to the similar conclusions since announced alike by Schützenberger and by Boutlerow, which require our farther consideration. Schützenberger has called attention to the fact that, besides the well recognized types of metallic oxyds, there are intermediate ones, in which the quantivalent ratios between the metal and the oxygen are less simple than in these. "The oxyds of tin, mercury, copper, lead, zinc, manganese, and iron, when prepared in different ways, vary considerably in their proportions of oxygen. Thus cupric oxyd from the calcined nitrate evolves oxygen when dissolved in nitric acid, and ferric oxyd got by a similar process, if we assume the formula to be $Fe_2O_3$, gives the value of $Fe = 54$; while for a similar oxyd from the calcined oxalate, if we admit the same formula, we get $Fe = 56$. In other words, 24 parts of oxygen are in the first case united with 54 and in the second with 56 parts of iron." (**18.**) Schützenberger has also obtained an oxyd of copper $Cu_5O_4$, and another, well crystallized, which is $Cu_3O_2$.

The studies of Haas, and those of Gibson and Morrison, have shown similar results from the action of hydrogen peroxyd on the ordinary oxyds of many metals. Thus the latter chemists not only got higher oxyds of zinc and of cadmium, but of magnesium and aluminium. The buff-colored zinc peroxyd (long since observed by Thénard), which, according to Haas, has the characters of an integral compound, and not of a mixture of oxyds, presents a composition corresponding either to $Zn_3O_5$ or $Zn_5O_8$. The higher oxyd of magnesium was found to be $Mg_5O_6$, and the brown peroxyd of cadmium $Cd_4O_7$; while manganous oxyd gave, by the action of hydrogen peroxyd, $Mn_2O_3$ or $Mn_5O_8$.[1] The late remarkable studies by Franke of this metal have, moreover, yielded him, besides a crystalline oxyd, $Mn_5O_8$, not only a volatile oxyd, $MnO_3$, but a blue-colored gaseous oxyd, $MnO_4$, apparently corresponding to osmic oxyd.[2] Spring and De Boeck have, moreover, recently obtained a soluble and colloidal hydrate of an oxyd, $Mn_7O_{12}$.[3]

[1] Gibson and Morrison, Proc. Royal Soc. Edinburgh, cxix. 146.  Haas, Ber. Deutsches Chem. Gesellschaft, xvii. 2249.
[2] Franke, Jour. für Prakt. Chemie, 1887, xiv. 166.
[3] Bull. Soc. Chimique de Paris, xlviii. 172.

§ 104. The quantivalent ratios thus observed lead to the conclusion that, as in the case of compounds of carbon and hydrogen, which present analogous variations, we have to deal with bodies of complex structure and high integral weights, belonging to one or more progressive series, the terms of which may have a common difference of $M_2$ or of O. We shall then have not only $x(M_2O)$, $x(MO)$, $x(M_3O_4)$, $x(M_2O_3)$, $x(MO_2)$, $x(M_2O_5)$ $x(MO_3)$, $x(M_2O_7)$, $x(MO_4)$, but intermediate terms, so that, if in the first term $x = 50$, these various oxyds, though including several homologous series, might all be represented by a general formula, $M_{100}O_{50} + nO$. The coefficient of M, for highly condensed solid species like periclase, magnetite, hematite, polianite, quartz, and cassiterite, will, however, in accordance with the law of density already set forth, approach 2000.

If for these various metallic oxyds we make the oxygen $= O_{100}$, we shall have for M the numbers, 25, 28.5, 33.3, 40, 50, 57, 58.3, 60, 62.5, 66.6, 75, 83.3, 100, 125, 150, 200. The existence of a similar series is, moreover, well illustrated in the cases of native sulphids and arsenids.

Taking of the former only the recognized species and their formulas as admitted by Groth,[1] we find, by making the sulphur $= S_{100}$, the following values for M :— 50, 57.1, 66.6, 70, 71.4, 77.2, 80, 83.3, 87.5, 100, 120, 125, 133.3 — 600; the latter being for a crystallized matte of iron and nickel, lately described by J. B. Mackintosh.[2] For the native arsenids, in like manner, with $As_{100}$, we have for the values of M the numbers, 33.3, 50, 66.6, 100, 150, 300, 500. These very varied ratios for the oxyds and sulphids, being deduced from arbitrary formulas, are, doubtless, in some cases, but approximations.

§ 105. The highest numbers observed for the metals in the above series of their compounds with oxygen, sulphur, and arsenic are those which have generally been assigned as the valencies or atomicities of these elements, it having been assumed by some chemists that the relations thus designated correspond to fundamental and invariable properties of the elements named, whereby a limit is fixed to

[1] Tab. Übersichte der Mineralien, Zweite Auflage, 1882.
[2] Trans. Amer. Institute Mining Engineers, 1887.

their power of forming combinations. The admission of its variability, which it is sought to explain by the expedient of free and latent bonds of affinity, is, however, of itself, as Mendeléef has remarked, a reason for rejecting this hypothesis of valency. It is not the capacity for saturation of any given element which we have to consider in a chemical compound, in which, on the contrary, "the entire system is maintained by forces which belong to each individual particle."[1] In other words, the chemical species is an integer which remains undisturbed under the ordinary conditions of temperature and pressure. There is no reason *à priori* against the possible existence of a metallic sulphid in which the quantivalent ratio of metal to sulphur shall be greater than $6:1$, or of a corresponding oxyd in which it shall be greater than $2:1$; that is to say, sulphur may be octovalent and oxygen tetravalent or hexavalent. The power of some melted metals, as copper, to hold in solution portions of oxyd, and of silver while in fusion to retain dissolved, that

[1] Mendeléef, The Periodic Law of Chemical Elements, Chemical News, xli. 108.

is to say, in combination, considerable quantities of oxygen, shows the existence, at least under certain conditions of temperature, of such suboxyds, which at lower temperatures are disintegrated. The separation of metallic copper in a filamentous form during the cooling of rich copper mattes is evidence of such a dissociation of an unstable sulphid; and the present writer long ago pointed out a similar separation, during the cooling of a ferriferous copper matte, of a portion of metallic iron, which, together with magnetic oxyd, remains diffused through the crystalline sulphid.[1]

§ 106. If, now, as we have done for the oxyds, sulphids, and arsenids, we take all the known hydrocarbons, and from their centesimal com-

[1] A matte got from the first fusion of a roasted cupriferous pyrites held 45 per cent of copper, with a little zinc, and was strongly magnetic. Oxydized by nitric acid or by bromine, it left more than 10 per cent of pure magnetic oxyd of iron. It, moreover, precipitated freely metallic copper and lead from solutions of these metals, and gave up to dilute acids the larger part of its iron, with evolution of free hydrogen and hydrogen sulphid gases, the latter apparently formed by the action of nascent hydrogen upon the metallic sulphid; from all which facts it was regarded as an intimate mixture of metallic iron, magnetite, and sulphids. Trans. Amer. Assoc. Adv. Science, 1873, p. 143.

position calculate formulas with $C_{100}$, we shall obtain ratios of C : H as varied as those which we have found for the compounds of M with O, S, and As. In the case of these metallic compounds, for the most part fixed and insoluble, we have not the means of directly determining their integral weights or their chemical relations. For the hydrocarbons, on the contrary, the centesimal composition, the vapor-density, and the genetic relations of each — that is to say, the modes of its production and its various derivatives — give for the hydrocarbon in question such a chemical history that we are enabled to fix its integral weight, and to assign it to one or another of the great homologous series. From the genealogical history of each such species, we are even enabled to devise structural formulas which serve to explain its origin and the results which it may yield by metagenesis. The fixed and insoluble metallic compounds, on the other hand, can, for the most part, be studied chemically only by the aid of processes which cause their more or less complete heterogeneous disintegration.

The numbers expressing the quantivalent

ratios of these compounds show that we have to deal with high integral weights, and with much less simple proportions than those observed in the volatile oxyds, sulphids, and chlorids. It was from the consideration of these and similar facts that, as we have seen (§ 103), J. P. Cooke and subsequently Schützenberger and Boutlerow were led to doubt the absoluteness of the law of definite proportions. Boutlerow conceives "that the combining-weight of carbon may vary from 12.0 to 11.8, and supposes a change in the chemical value of that element; that is to say, the amount of carbon united with 32 parts of oxygen may vary from 12.0 to 11.8 parts, the resulting compounds, though not identical, being similar in chemical properties. Schützenberger, in like manner, admits the variability in value on either side of a point of maximum stability, which is in most cases attained. He concludes that 'all of these results lead directly to the conclusion that the law of definite proportions is not rigorously exact, unless we are willing to admit in each particular case the existence of compounds more or less oxygenized than those hitherto known to us, which may occur mixed with the products of the reaction."[1] This suggested explanation of the facts, which, however, does not appear to be acceptable to Schützenberger, Boutlerow, or Cooke (who has published a valuable discussion of the subject), is, nevertheless, an approximation to what we conceive to be the true one, believing, however, that the oxyds with more or less than the ordinary proportions of oxygen are not necessarily, or even probably (with some apparent exceptions), admixtures, but definite intermediate oxyds, being members of great progressive series."

§ 107. "The conception that these variations in composition are due to the presence of ad-

[1] See, for the papers of Schützenberger and Boutlerow, Bull. de la Société Chimique de Paris, 1883, xxxix. 257-263, and farther Boutlerow, Jour. Russ. Chem. Society, 1882, pp. 208-212, in abstract in Jour. Chem. Soc. London, xlii. 922; an extended analysis of them is in the American Journal of Science, in the same year, xxvi. 63. The results obtained, according to Schützenberger, " conduisent directement à cette conclusion, que la loi des proportions définies n'est pas rigoreuse, à moins qu'on ne veuille admettre dans chaque cas particulier l'existence de composés plus oxigénés ou moins oxigénés que ceux connus jusqu'à présent, et qui se trouveraient en mélange avec le produit principal de la réaction."

mixed portions of more and less oxygenized compounds is not acceptable to Schützenberger, for the evident reason that it becomes inadmissible when we have to deal with gaseous or vaporous species, such as carbon dioxyd and as water-vapor. This, as generated by passing hydrogen gas over ignited copper oxyd, presents, according to him, variations in the proportions of H : O, of from 1.00 : 7.95 to 1.00 : 8.15 by weight; that corresponding rigorously to the volumetric relation, 2 : 1, being very nearly 1.00 : 7.98. Water thus obtained with an excess of oxygen, though neutral, possesses, according to Schützenberger, oxydizing powers; the same being true of carbon dioxyd with the larger proportion of oxygen."

§ 108. "Professor Cooke had already, in 1860, arrived at a conclusion similar to that subsequently reached by Schützenberger and by Boutlerow; namely, that 'the law of definite proportions is not so absolute as has been hitherto supposed,' and that the chemical value of the elements may change within certain limits. In his discussion, in 1883, of these views, as enunciated by the chemists just named, Cooke remarks, — 'Such opinions are certainly very revolutionary, and, if they prevail, must entirely change the fundamental conceptions of chemical philosophy. Chemical combination can no longer be regarded as the juxtaposition of the definite, invariable masses which we call atoms, but must be considered as the "reciprocal saturation" or "interpenetration" of masses, which may vary with the relative strength of their chemical energy acting at the time; and this change of the fundamental conception is inconsistent with the atomic theory, and with the superstructure which modern chemistry has built upon it.' Cooke adds that, while holding that 'the atomic theory is the only basis on which a consistent philosophy can at present be built,' 'he is rather drawn to that view of nature which refers all differences between substances to dynamical causes, and which regards the atomic theory as only a temporary expedient for representing the facts of chemistry to the mind.'"[1]

§ 109. "It has been shown that the writer does not share the interpretation of these vari-

[1] American Journal of Science [1883], xxvi. 310, 316.

ations which is given by Cooke, Schützenberger, and Boutlerow, but regards, as was already said by him in 1867, the facts 'from which some have suggested a deviation from the law of definite proportions' as 'only an expression of that law in a higher form'; believing that, so far as we yet know, the laws of measure, number, and weight in chemistry are invariable. To one who, since 1853, has persistently combated the atomic hypothesis as contrary to a sound philosophy, and throughout all the succeeding years has sought to build up without it a new theory of chemistry upon a dynamic basis, it is, however, no small satisfaction to find Professor Cooke, who has been among the ablest defenders of this famous hypothesis, is at last led to look upon it as 'only a temporary expedient.'" (**18.**)

§ 110. The question now arises, whether it is possible to admit the existence in a gaseous form, corresponding to ordinary water-vapor, of a species having the composition found by Schützenberger for water with an excess of oxygen (§ 107). From the analogy of the oxyds already noticed, we may suppose other compounds of oxygen and hydrogen than the two hitherto recognized, so that, outside of "the point of maximum stability" represented by ordinary water, there may be, for example, an oxyd $H_{100}O_{50} + O$. We may farther suppose such a union to give rise to a condensed vaporous species, the integral weight of which should be represented by this formula, or by some multiple thereof, corresponding to the very dense vapor of water above its critical point, which, as we shall endeavor to show further on, is such a condensed species having the ordinary quantivalent ratio of 100 : 50, while that of the more oxygenized compound is 100 : 51. If, now, both of these undergo a similar homogeneous disintegration, — as in the case of hexad sulphur, tetrad fluorid of hydrogen, or dyad iodine-vapor, — we shall find that, "while ordinary water-vapor, with the ratio for hydrogen and oxygen of 2 : 1 by volume, and of 1.0 : 7.9816 by weight, equals $(H_{100}O_{50}) \div 50$, the ratio 1.0 : 8.142 (or very nearly that observed by Schützenberger) corresponds to $(H_{100}O_{51}) \div 50$. These figures, arbitrarily assumed, will serve for illustration; but it may be remarked that the integral weight for the gaseous species, $H_{100}O_{51} = 898.16$, which may be supposed to have a momentary existence before homogeneous disintegration, is not much greater than the weights observed for the vapors of stannic and aluminic iodids — $SnI_4 = 534.1$, and $Al_2I_6 = 813.2$" and very much less than the dense vapors yet to be noticed.

§ 111. The principle of chemical condensation and expansion is universal; that is to say, there is, theoretically, no limit to the process of integration, nor in like manner to that of disintegration, until we attain a supposed *materia prima*. The latter process, as is known, takes place in elemental species, as well as in the gases and vapors of compounds, such as fluorhydric acid, nitric peroxyd, acetic acid, paraldehyde, and turpentine-oil. Under the hitherto known conditions of experiment, hydrogen gas, the lightest known species, which is the type of the normal chemical integer, remains chemically unchanged by heat, and, "the weight of two portions of this element ($H_2 = 2.0$) being the unit, the weight of a like volume of any other gas or vapor is its equivalent weight;

that is to say, the weight of a volume equal to $H_2$. This, in the language of the atomic hypothesis, is its molecular weight; the so-called atomic weights being the smallest combining-weights of the elemental species, compared with $H = 1.0$."

§ 112. "The normal integers of oxygen, hydrogen, and chlorine, and of bromine and iodine vapors, are thus double or dyad integers, while ozone and the vapor of selenium below 800° are triple or triad, and that of sulphur-vapor below 550° is sextuple or hexad; all of these, however, at higher temperatures, assuming the densities of the normal or dyad integers. The quadruple integers of phosphorus and arsenic vapors, according to Mensching and Victor Meyer, undergo a like change, while that of antimony is either dyad or monad, and the double integer of iodine, at about 1500°, becomes, as is well known, a single or monad integer, a result which, at the highest temperatures of experiment, is already partially attained for bromine and chlorine; the vapors of mercury and cadmium being known to us only in this condition of monad integers." (**18.**)

The subject of homogeneous disintegration by heat is farther illustrated by the changes of the various compound species named above. "These integers of varying densities, and of varying equivalent weights, alike of so-called elements and of compounds having a similar centesimal composition, may be conveniently designated higher and lower integers; the terms implying at the same time higher and lower specific gravities, and higher and lower equivalent or integral weights."

"The chemical species, to those agencies which do not effect its disintegration, is a complete entity, unit, or integer. This, in the case of homogeneous integration in gases and vapors, is generated by the condensation into a single volume of two or more volumes of a less dense species. The designation of polymers given to such condensed species, implying that they are made up of many parts, carries with it the notion of building by additions, and thus of complexity rather than of integrity: it is therefore rejected." (**18.**) The specific gravity of a gaseous or vaporous species, at standard temperature and pressure, varies directly as its equivalent or combining weight. "Since this weight is nothing else than the specific gravity of the gas or vapor, hydrogen being unity ($H_2 = 2.0$), it may, dispensing with all hypothesis, be designated as the weight of the integer: or, in other words, as the integral weight of the gaseous or vaporous species." **(19.)**

§ 113. But, while the process of disintegration is theoretically limited only by the term of one or more species of highly attenuated matter, and while there is reason to believe that this process attains a far greater extension in stellar worlds than on the earth, where its operation is limited by the conditions of temperature and pressure, the question arises, what are its actual limits under present conditions? The dyad integer, as seen in hydrogen and chlorine gases, might, but for the evidences of farther dissociation in elemental species just noticed, be supposed to mark that limit. That a similar process of homogeneous disintegration comes into play in complex species was long since insisted upon by the author. In ordinary cases of combination between unlike gaseous species, as when oxygen and hydrogen unite to form water-vapor, there is condensation, the volume of the combining oxygen disappearing in that of the hydrogen. In the production of chlorhydric gas from chlorine and hydrogen, however, no such condensation is observed, the volume of the product being the sum of those of its constituents. Hence, as pointed out in 1853 (§ 4), we conclude that "a union takes place, followed by immediate expansion without specific difference," so that we are not able to "observe the intermediate stage." The compound of the dyad integer of hydrogen ($H_2$) with a similar integer of chlorine ($Cl_2$) is at once dissociated into two dyad integers of chlorhydric gas, $H_2Cl_2 = 2HCl$ — the double process of heterogeneous integration and homogeneous disintegration taking place without a perceptible interval of time, so that no change of volume is observed. But, although the process of formation of chlorhydric gas cannot be arrested at this intermediate stage, or, in other words, although the normal species, $H_2Cl_2$, is still unknown to us, the corresponding species in the case of fluorhydric acid has been discovered. The compound of fluorine with hydrogen (which, as Moissan has found, is directly formed by spontaneous union, with explosion, even in the dark) has, as was shown by Gore, a density, at and near 100°, corresponding to HF, or twenty times the weight of hydrogen. Mallet, however, observed, in 1881, that the density of fluorhydric vapor at 30° is double this, or forty times that of hydrogen, corresponding to $H_2F_2$, although at higher temperatures, as he has remarked, "dissociation manifestly takes place, leading to the production of diatomic molecules of HF."[1]

§ 114. It was maintained by the present writer, in 1848, that allotropism in elemental bodies is identical with the so-called polymerism in hydrocarbonaceous compounds, and is connected with greater or less degrees of condensation, which are manifested by differences in specific gravity, alike among gaseous and among solid species, as set forth in § 5, in extracts from papers then published. Subsequently, in 1887, recalling this same subject of homogeneous integration or polymerism, and

[1] American Journal of Chemistry, iii. 189.

the great additions which since 1848 have been made to our knowledge of such cases, it was said, after referring to the chemical instability of some gaseous and volatile species, that "there are, doubtless, other polymers which, like tri-sulphur, $S_6$ [the hexad sulphur-vapor], and di-pentine, $2(C_5H_8)$, exist within certain limits of temperature and pressure. In this connection, the studies of Cagniard de Latour, of Drion, and of Andrews, on the conversion of liquids into gases, are very important, and help to enlarge our conceptions of this polymerism in vapors under great pressure" (§ 53).

§ 115. The experiments of Cagniard de Latour, published in 1822–23,[1] showed that ether, alcohol, naphtha, and even water, when heated sufficiently, in hermetically sealed tubes, appear to be wholly converted into vapor, in spaces occupying from two to four times the original volume of the liquids. These observations were, in 1859, confirmed for ether (ethyl oxyd) by Drion,[2] who also got similar results for ethyl chlorid, and, moreover, found the coefficient of

[1] Ann. de Chim. et de Phys. (2), xxi. 127, 178; xxii. 410.
[2] *Ibid.* (3), lvi. 33.

dilatation to increase very rapidly with the rise of temperature of the liquid. Ethyl chlorid is at 130° one and a half times as dilatable as air, and at 170° passes into a dense vapor, as in the experiments of Cagniard de Latour; in which, as in those of Drion, the tension of the vapors at various temperatures was determined. Thilorier had already noted that the dilatability of liquid carbon dioxyd at temperatures between 0° and 30° is very much greater than that of air, and Drion, from his own observations, conceived that this high and augmenting rate of expansion is general; a fact since confirmed by the studies of many others, which show that the coefficient of dilatation in liquids increases rapidly with the rise of temperature under pressure, and soon exceeds greatly the constant coefficient of air.

§ 116. Thomas Andrews, in 1861, undertook the study of carbon dioxyd, for the liquid form of which, also, he observed the passage into a gaseous state under pressure, and found that above 30.92° this body remains gaseous even under a pressure of 300 atmospheres. Below this temperature, however, the dense gas passes again into the liquid state without any sudden transition. These observations, having been communicated to W. Allen Miller, were discussed by him in connection with the previous ones of Cagniard de Latour, Thilorier, and Drion, in the third edition of his "Chemical Physics," in 1863. Therein he considers the general fact of the passage from a liquid to a gaseous state without the usual great augmentation of volume, or, in other words, the production, at elevated temperatures, under pressure, of very dense vapors, and concludes that "there exists for every liquid a temperature at which no amount of pressure is sufficient to retain it in a liquid form." From this it would follow that for the liquefaction of many species, such as oxygen, nitrogen, and hydrogen, there would be required, not only pressure, but great artificial cold, by the combined action of which the liquefaction of these hitherto permanent gases has since been attained.

§ 117. Miller, at the same time, called attention to the fact that the dense vapors of ether and alcohol, discovered by Cagniard de Latour, have not, at the temperature of formation, the elastic force which should belong to the normal vapor of these bodies at such temperatures. The pressure of the dense vapor of ether at 188° was, according to Cagniard de Latour, equal to 37.5 atmospheres, and that of alcohol at 258.7° to 119 atmospheres, " whereas, if Marriotte's law held good in these cases, calculating from the volume of vapor which a certain volume of each liquid yields under the atmospheric pressure, ether should have exerted a force equal to about 157 atmospheres, and alcohol at least 318 atmospheres." Miller also notes that when, after these bodies have once passed into the state of dense vapors, the temperature is farther raised, the elastic force of the vapor increases very much more rapidly than that of air under similar conditions. Air, under a pressure of 37.5 atmospheres, at 188°, would, if raised to 250°, exert a force of 42.4 atmospheres, and at 325° one of 48.6 atmospheres; while the pressures for the dense ether-vapor were, at these temperatures, respectively, 86.3 and 130.9 atmospheres.

From the facts here shown, it would appear (1) that such a volatile liquid, under pressure, passes into a vapor of great density, which, near its point of formation, has a tension much less than should belong to the normal vapor of the liquid in question ; (2) that this abnormally dense vapor, on augmentation of temperature, not only shows a much greater rate of expansion than air or water-vapor, but that its rate, instead of being, as for these, constant, shows a rapidly increasing coefficient of dilatation. It will be seen that these phenomena are in accord with what is observed in sulphur; this, near its point of condensation, yields a dense hexad vapor, which, at a higher temperature, is disintegrated into a less dense dyad vapor. The similar changes in the vapor of iodine and of fluorhydric acid are other examples of the same kind.

§ 118. The farther results of the remarkable studies of Andrews, in this field, are embodied in his two Bakerian lectures before the Royal Society of London, 1869 and 1876, the first being entitled, The Continuity of the Gaseous and Liquid States of Matter; and the second, The Gaseous State of Matter.[1] The ideal or

---

[1] Philos. Trans., 1869, part i. 575-589; 1876, part i. 421-449.

perfect gaseous state is nearly attained in gases at ordinary pressures and at temperatures greatly above their points of liquefaction, in which case their deviations from this state can only be discovered by careful experiment. In such a condition, matter would implicitly obey the forces which act upon it from without; the volume being inversely as the pressure applied, and the coefficient of expansion by heat being constant. In ordinary gases and vapors there are, however, in the language of Andrews, "two distinct causes of internal disturbance, whose results are directly opposed; and, according to the nature of the gas and the conditions of temperature and pressure, sometimes the one and sometimes the other predominates. One of these disturbances is due to internal forces tending to produce a diminution of volume; the other is due to molecular conditions producing a resistance to diminution of volume other than that which occurs in a perfect gas, . . . whereby the gas undergoes a less diminution of volume than would occur in the case of an ideal gas obeying Boyle's law." In the case of ordinary liquids, the resistance to change of volume from increased pressure is very great; in such liquids as carbon dioxyd it is not so great, but augments as the pressure is increased. "The resistance of liquids and gases to external pressure tending to produce a diminution of volume, proves the existence of an internal force of an expansive or resisting character. On the other hand, the sudden diminution of volume without the application of additional pressure externally, which occurs when a gas is compressed, at any temperature below its critical point, to the volume at which liquefaction begins, can scarcely be explained without assuming that a molecular force of great attractive power comes into operation, and overcomes the resistance to the diminution of volume which commonly requires the application of external force."

§ 119. The observations cited in § 116 show that the distinction between the gaseous and the liquid states is one not clearly defined. In the case of carbon dioxyd, according to Andrews, "by varying the pressure or the temperature, but always keeping the latter above 30.92°, the great changes of density which occur about this point produce flickering movements, resembling, in an exaggerated form, the appearances due to the mingling of liquids of different densities, or of columns of hot and cold air. ... The ordinary gaseous and ordinary liquid states are only widely separated forms of the same condition of matter, and may be made to pass into one another by a series of gradations so gentle that the passage shall nowhere present any interruption or break of continuity. They are only distant stages of a long series of continuous physical changes." "Under certain conditions of temperature and pressure carbon dioxyd finds itself, it is true, in what may be described as a state of instability, and suddenly passes, with the evolution of heat, and without the application of additional pressure, or change of temperature, to the volume which, by the continuous process, can only be reached by a long and circuitous route. In the abrupt change which here occurs, a marked difference is exhibited, while the process is going on, between the optical and other physical characters of the carbon dioxyd which has collapsed into the smaller volume, and that not yet altered." Of carbon dioxyd at 35.5°, and 108 atmospheres, Andrews remarks that, while it is reduced to $\frac{1}{430}$ of the bulk which it occupied at one atmosphere, "it is impossible to say whether it is gas or liquid. We have no valid grounds for assigning it to one condition or the other."

Similar phenomena were observed by Andrews for nitrogen monoxyd, chlorhydric acid, ammonia, carbon disulphid, and ethyl oxyd, all of which exhibit, at fixed pressures and temperatures, critical points and rapid changes of volume, with flickering movements. "A vapor may be defined to be a gas at any temperature under its critical point."

Sajotschewsky, in 1879,[1] published the results of a series of similar studies of many liquids, and found the critical point of ether to be 190° at a pressure of 36.9 atmospheres, and that of alcohol 234.3° at a pressure of 62.1 atmospheres.

§ 120. The studies of dense vapors by J. B. Hannay and James Hogarth,— On the Solubility of Solids in Gases, — published in 1880, constitute an important contribution to the

[1] Beiblätter, iii. 752.

subject.[1] They found the critical point of pure alcohol, as a mean of many experiments, to be 234.6°, with a pressure of 65 atmospheres. Extending their experiments, they observed that a saturated solution of potassium iodid in forty parts of alcohol passed integrally, a little above the critical point of pure alcohol, into a gaseous state; and, farther, that alcohol-vapor, at a temperature of 300°, reduced by pressure to the volume which it had occupied when still liquid, completely dissolved a crystal of potassium iodid. Similar results were got with solutions of potassium bromid, and of cobaltous and ferric chlorids, the vapors of these latter retaining their characteristic blue and yellow colors; though a gradual change, probably due to formation of ethyl chlorid, took place in the latter, as well as in a solution of ferric chlorid in ether. The solution of anhydrous calcium chlorid in alcohol was uninterrupted in the passage through the critical state, though the solution at 230° separated into a lighter and a denser layer, both of which, on farther eleva-

[1] Proc. Roy. Soc. London, xxix. 324, and Chemical News, xli. 103-106.

tion of temperature, passed successively the critical point at 250° and 255°, the denser liquid reappearing on reduction of pressure; showing, as remarked by the authors, a distinct combination of the chlorid with part of the alcohol. Similar phenomena were observed with the solutions of ferric chlorid. The absorptive spectra of alcoholic solutions of cobaltous chlorid, and of chlorophyll, remained unchanged above the critical point. A solution of sulphur in carbon disulphid also passed unchanged the critical point, and sodium appeared to dissolve in hydrogen gas at a high temperature. These experiments, in the words of their authors, give "a farther proof of the perfect continuity of the liquid and the gaseous states, and also a complete proof of the solubility of solids in gases."

§ 121. Water has also its critical point, which, according to Cagniard de Latour, is near that of melting zinc, or not far from 400°, when it passes into about four volumes of dense vapor; but the high temperature, and the action of the liquid on glass (although partially prevented by the addition of sodium carbonate), render experiments with it difficult. Our authors, however, observed that water, above its critical point, takes into solution silica, alumina, and zinc oxyd, and, by farther experiments, not yet clearly explained, apparently found, under great pressure at high temperatures, a gaseous solvent for carbon, which they succeeded in obtaining in crystals having the chemical and physical characters of the diamond.[1]

It is difficult to over-estimate the importance, in physiological mineralogy and in geogeny, of these facts as to the solvent powers of gases and vapors, and of the not less significant union of gases and vapors with liquid and solid forms of matter. Hannay has shown [2] that glass at 200°, and under a pressure of 200 atmospheres, absorbs large quantities both of oxygen and of carbon dioxyd gases, which, when cooled under pressure, remain permanently fixed. These are, in great part, slowly given off again at 300°, and, when the glass is rapidly raised to the fusing-point, are suddenly expelled, converting it into a frothy mass. In this connec-

---

[1] Hannay and Hogarth, Chemical News, xli. 106; also, N. S. Maskelyne, *ibid.*, 97.

[2] *Ibid.*, xliv. 3.

tion should be noted the early observations of Spallanzani, and the later ones of Damour and Boussingault,[1] on the transformation of obsidian or volcanic glass into pumice under similar circumstances, from the loss of volatile matters, in part water and in part nitrogen, with some chlorhydric acid; as also the essay of Charles Sainte Claire Deville on the phenomena of trachytism,[2] and the late studies of Tilden and Shenstone on the union of various salts with water at high temperatures (§ 73), with the present author's discussion of the subject.[3]

§ 122. In this connection, it should be stated that the experiments of Cagniard de Latour on alcohol were made with a spirit of specific gravity 0.844, which corresponds to an admixture of about 18.0 per cent of water. Messrs. William Ramsay and Sidney Young[4] have recently assigned for the critical point of pure alcohol a temperature of about 243°, when

[1] Ann. de Chim. et de Phys. [4], xxix. 543.
[2] Sur le trachytisme des roches; Comptes Rendus de l'Académie des Sciences, xlviii. 15–23.
[3] Mineral Physiology and Physiography, 220–222.
[4] "On the Thermal Properties of Alcohol," Philos. Trans., 1886, part i.

they found the distinction between gas and liquid to disappear at a pressure of nearly 47,700 mm., or not quite 63 atmospheres. The gramme of alcohol, which has at 4° a volume of 1.2403 c.c., attains at the critical point a volume of about 3.5 c.c., so that the specific gravity of the dense vapor is nearly 0.28, water being 1.00.

In a more recent paper, on the Nature of Liquids,[1] Messrs. Ramsay and Young raise the inquiry why the law of Boyle, that the volume of the gas is inversely as the pressure, the temperature being constant, and the law of Gay-Lussac, that the volume of the gas increases by a constant fraction of its volume at 0° for each rise of 1°, — which obtain in the case of perfect gases, — "do not hold either near the condensing-points of gases or at high pressures." And they ask whether the abnormal density of the vapors which we have considered is "in any degree due to chemical association of molecules." "At any temperature below the critical one, when the volume of the gas is decreased, pressure rises until a

[1] L., E., and D. Philos. Magazine, February, 1887.

certain maximum is reached, when it becomes constant, and change of state occurs." The query is then propounded by our authors whether "the liquid condition thus reached is a purely physical one, or, as advanced by Naumann and others, there is in this case formation of molecular groups of definite complexity." Such a view, it will be remembered, was mentioned with approval by Playfair and Wanklyn,[1] in 1861, and was sustained by Alex. Naumann in his studies of the vapor-density of acetic acid, in 1870.[2] The view defended by Messrs. Ramsay and Young is, on the contrary, that "the molecules of stable liquids [that is, of liquids not undergoing heterogeneous disintegration by heat] are not more complex than those of gases," and that the anomalies in question are due to physical rather than to chemical forces.

§ 123. The authors, at the outset, are fettered by their proposition that "the fundamental concept in chemistry, as well as in physics, is the molecular and atomic constitu-

---

[1] Trans. Roy. Soc. Edin., xxii.
[2] Annal. der Chem. und Pharm., clv. 325.

tion of matter," and their conclusions appear to be vitiated by the failure, which we have elsewhere noticed in others, to recognize the limits between physics and chemistry (§§ 11, 97, 124). The perfect gaseous state of matter exists only within certain limits of temperature and pressure, within which it is wholly obedient to these external influences, which affect its volume in a regular and constant manner. The study of such changes belongs to the dynamical history or the physics of matter. The action of these external influences is, however, modified by what Andrews designated as two causes of internal disturbance, the one conducing to condensation and the other resisting it; or, in other words, by "that tendency in nature which constantly leads to unity, condensation, identification" (§ 14), and its antagonist. "In the capacity of such changes consists the chemical activity of matter" (§ 3), and all such internal disturbances, while they are attended by new dynamical phenomena, and are favored by changes of temperature and pressure, and by radiant energy, are manifestations of chemism. The internal changes which, in the cases discussed by Ramsay and Young, disturb the regular action of the dynamic laws of temperature and pressure in gases or vapors near their condensing-points, or at high pressures, do not differ from those changes which occur, under similar conditions, in the vapors of sulphur, of iodine, or of fluorhydric acid, and cannot, we think, in any sense, be excluded from the domain of chemical phenomena.

The doctrine referred to by Messrs. Ramsay and Young as having been advanced by Naumann and others, that the explanation of these deviations from the dynamical laws of temperature and pressure in dense vapors and in liquids "are due to the formation of molecular groups of definite complexity," is, when stated without the use of the molecular hypothesis, the one which the present writer has persistently maintained since 1853, having reiterated it in 1867, and often subsequently (§§ 20–22). Its application to liquids and solids, as well as to denser gases and vapors, appears to him to give the only rational explanation of the facts of chemistry. For liquids, as for gases, within a certain range of temperature, the law of thermic expansion is constant ; but above this, towards the critical point, their rapidly augmenting rate of expansion, like the similar expansion of water below 4°, indicates the production of less condensed species with lower integral weights, which, above the critical point, pass into dense vapors.

§ 124. We have elsewhere noticed (§§ 85-88) the identification of the processes of vaporization and chemical dissociation, as insisted upon by Henri Sainte-Claire Deville in 1857-65. They were subsequently "well resumed in his final statement in 1873, — that Isambert, in his studies of the compounds of ammonia with chlorids, 'not only demonstrates the analogy between the phenomena of dissociation and of vaporization, but establishes their complete parallelism.' Hence, in the opinion of Deville, 'there is, according to these ideas, no essential difference between physical phenomena and chemical phenomena, or rather the passage from the one to the other is continuous.'"[1]

[1] "Isambert . . . non seulement démontre l'analogie entre les phénomènes de dissociation et de vaporization, mais en

It may be said, in comment upon this remarkable statement, that, until the true nature of the processes of volatilization and vaporous condensation had been established, those chemical phenomena marked by changes of state had been wrongly regarded as physical (or dynamical) phenomena. By 'change of state' are here meant all changes between the conditions of gas, liquid, and solid, as also the transformations of these; which are always marked by alterations in density, as well as in other physical characters. Illustrations of these changes are familiar alike in simple and in compound species."[1] (**18.**)

établit le parallélisme complet. . . . Il n'y a aucune différence, d'après ces idées, entre les phénomènes physiques et les phénomènes chimiques, ou plutôt le passage des uns aux autres se fait par des variations continues." Report by Deville on a memoir by Troost and Hautefeuille, Sur les transformations isomériques et allotropiques; Comptes Rendus de l'Académie des Sciences, 1873, lxxvi.

[1] Of phosphorus, which affords perhaps the most instructive example of these changes, there are known: (1) The vaporous tetrad species, which, according to Victor Meyer, gives indication of conversion into a more elemental species at very high temperatures; (2) liquid phosphorus, which boils at about 279°,—when it has a density of 1.5285,—and may, under proper conditions, be cooled to 20°, or even to 0°, but below 44° is readily changed, with rise of temperature, into the solid,

§ 125. Having dealt with the question of the relations of pressure to gases and vapors, we now proceed to consider it with regard to liquids and solids, and notice, in the first place, the observations of H. C. Sorby, as set forth in his Bakerian lecture before the Royal Society of London in 1863, entitled, The Direct Correlation of Mechanical and Chemical Forces.[1] Therein, after calling attention to the results of Bunsen and of Hopkins, who showed that bodies which expand in passing from the solid to the liquid state have the point of fusion

transparent, colorless, crystalline species, (3) which is luminous in the air, soluble, poisonous, a non-conductor of electricity, having at 40° a density of 1.806. This is converted by sunlight, by the action of iodine, by a temperature of 230°, or more rapidly, with evolution of heat from the condensation, when exposed to the temperature of boiling sulphur [450°], into (4) the amorphous red species, insoluble, non-luminous, and non-poisonous, with a density of about 2.10. This last species volatilizes *in vacuo*, without fusion, at 550°, and is redeposited in red crystals, which are perhaps identical with the red crystallized metalloidal species (5) got by melting lead with phosphorus under pressure, which has a density of 2.34 at 15°, and is a conductor of electricity. The forms of red phosphorus with intermediate densities are possibly mixtures of the last two species.

[1] Proc. Royal Society London, xii.

raised by pressure, and to the observation of Sir William Thomson, that ice, which contracts in melting, has, on the contrary, its fusing-point lowered by pressure, he adds that, "when ice melts and mixes with water, it may be looked upon as dissolving in it." Sorby then proceeds to consider the relations of pressure to the solution of other bodies in water, which he rightly regards as a chemical process (§ 12). In such solution, there is for the most part a condensation or loss of volume, which is often considerable. Reasoning from this, he was led to believe that pressure would in these cases increase the solvent power of the liquid, a fact which he was able to verify by experiments with pressures up to 100 atmospheres for sodium chlorid,[1] potassium sulphate, copper sulphate, and potassium ferrocyanid, and from the data obtained

---

[1] The present writer was afterwards enabled to confirm this by showing that a supersaturated solution of sodium chlorid formed at the bottom of a column of 1000 feet of brine (and thus under a pressure of 36 atmospheres), in a vertical boring at Goderich, Ontario, deposited, at the surface, upon the metal of the pump, regular cubes of the salt one-fourth of an inch in diameter, thereby obstructing the action of the pump. Chemical and Geological Essays, p. 204.

to calculate the mechanical equivalent of the force of solution. With ammonium chlorid, which dissolves with expansion, inasmuch as the solution occupies a greater volume than that of its constituents, Sorby found, on the contrary, that pressure decreased the solvent power of the water.

§ 126. The study of the chemical relations of great pressure to solids has since been studied with very important results by Walthère Spring of Liège. Chemical changes, whether of metamorphosis or metagenesis, may be divided into two distinct classes — those attended by augmentation, and those attended by diminution of volume. For the first class, the experiments of Cailletat and of Pfaff, among others, have shown that a pressure of from 60 to 120 atmospheres suffices to arrest the action of dilute sulphuric and nitric acids on zinc and on calcite; while a pressure of 40 atmospheres prevents the hydration of calcium sulphate. For the second class, in accordance with the observations of Sorby, pressure might be expected to favor chemical action; and this has been fully shown by the experiments of

Spring,[1] which were made in steel cylinders eight millimetres in internal diameter, in an apparatus permitting a measurable pressure of over 20,000 atmospheres *in vacuo*. Thus, copper filings and sulphur, mingled in proper proportions, and exposed to a pressure of 5000 atmospheres, at the ordinary temperature, are completely converted into crystalline cuprous sulphid, the excess of sulphur remaining disseminated in the mass; in this case, 138 volumes are condensed

---

[1] The following is a list of the principal papers by Spring on this subject:—

1. Recherches sur la propriété que possèdent les corps de se souder sous l'action de la pression, 1880, Bull. Acad. Roy. Belgique (2), xlix. No. 5; also, Ann. de Chim. et de Phys. (5), xxii. 170–217.

2. Formation de quelques arséniures métalliques par l'action de la pression, 1883, Bull. Acad. Roy. Belgique (3), v. No. 2.

3. Formation de sulfures métalliques sous l'action de la pression, 1883, *ibid.* (3), v. No. 4.

4. De l'action de la pression sur les corps solides en poudre. Réponse aux observations de MM. Ed. Jannetaz, Neel, et Clamont. *Ibid.*, Octobre, 1883.

5. Sur l'élasticité parfaite des corps solides chimiquement definés; Nouvelle analogie entre les solides et les liquides. Bull. Acad. Roy. Belgique, 1883 (3), vi. No. 11.

6. Sur les quantités de chaleur dégagées pendant la compression des corps solides, 1884. Bull. Soc. Chimique, xli. 488.

7. Sur les quantités de sulfures qui se forment par des compressions successives de leurs éléments, *ibid.*, xli. 492.

to 100. In like manner, other sulphids — such as those of zinc, cadmium, lead, silver, bismuth, and other metals — are readily produced by pressures up to 6500 atmospheres, as well as various arsenids; while Wood's fusible alloy of bismuth, tin, lead, and cadmium is at once generated by compressing in like manner the proper admixture of these metals in filings. In many instances, it is found necessary to reduce the first products to powder by filing, and to submit them to repeated compressions. Time is also an important element in the process of combination, and in the case of an admixture of silver and sulphur, as in others, it was found that the integration begun under pressure was continued after the pressure was removed; implying, as Spring has well observed, a veritable process of diffusion through the solid material, like that which takes place in liquids. In the reactions studied by him

8. Réaction du sulfate de barium et du carbonate de sodium sous l'influence de la pression, 1885. Bull. Acad. Roy. Belgique, (3), x. No. 8.

9. Réaction du carbonate de barium et du sulfate de sodium sous l'influence de la pression, 1886. Bull. Soc. Chimique, xliv. 166.

between potassium iodid and mercuric chlorid, and between barium sulphate and sodium carbonate, we notice examples, not only of metagenesis, but of double decomposition.

§ 127. Proceeding next to the results obtained in chemical metamorphosis by pressure, a first step, as Spring remarks, had already been made by showing the power of pressure to change ice below $0°$ to liquid water. He found, in his own experiments, that plastic sulphur and prismatic sulphur, by pressures of from 3000 to 6000 atmospheres, are readily converted into the denser and more fusible octahedral species. In like manner, amorphous arsenic is in considerable part changed into the crystalline and denser species, while red phosphorus appears to be partially converted into the metalloidal form. These reactions are comparable to the ready conversion by pressure of the gray, brittle species of tin, with a specific gravity of 6.0 or less, into the ordinary white, malleable form, with specific gravity 7.3 (§ 55). Prof. W. Chandler Roberts of the Royal Mint announced in 1882, to the Physical Society of London,[1]

[1] Chemical News, xlv. 431.

that he "had repeated and confirmed many of the experiments of Spring," among others the formation of fusible alloys.

All of these are examples of so-called allotropic modifications effected in solid bodies (§ 114). Most elemental solid species, however, do not appear to be susceptible of such modifications, and, in the case of such, pressure, even to 20,000 atmospheres, does not effect any permanent change in the specific gravity. In the ordinary metals and alloys, as lead, copper, steel, bronze, gold, and platinum, Spring has shown that the variations sometimes observed in density, especially for the last three named, are due to the presence of fissures or pores, which disappear under pressure, and that a constant density is thus attained, beyond which they resist obstinately any farther permanent diminution of their volume. Such bodies, he adds, are, however, not incompressible in the ordinary sense of the word, since, "through the whole duration of the pressure, it was easy to establish that their volume was diminished to a greater or less degree; but as soon as the pressure ceased, they regained their primitive state."

For any such body may be affirmed what he has said of platinum when reduced by great and repeated pressure to a specific gravity of about 21.45, that "it is really incapable of permanent compression, or, in other words, is perfectly elastic." These solids, then, share the slight compressibility and the perfect elasticity of liquids such as mercury and water.

§ 128. The filings of many metals, as lead, bismuth, tin, zinc, aluminium, and copper, under pressures of from 2000 to 6000 atmospheres, become completely welded or soldered together, and even assume the crystalline texture which belongs to fused masses. Lead is thus aggregated at 2000 atmospheres, and at 5000 atmospheres flows like a liquid into all the joints of the compressing apparatus, while tin welds at 3000 and flows at 7500 atmospheres. These studies of Spring thus confirm the earlier observations of Tresca on the flow of metals in the cold, under great pressure, and show that the distinction between liquids and solids is one of degree. Similar results as to the conversion of powders into crystalline solids were obtained with manganese dioxyd, and also with artificially prepared sulphids of zinc, lead, and arsenic, that of zinc becoming transparent. Various salts were transformed, under pressures up to 5000 atmospheres, into crystalline, transparent masses, and in some cases, as with potassium bromid and the hydrated sulphate, hyposulphite, carbonate, and acetate of sodium, the phenomena of liquidity and flow were observed. In a subsequent study of many salts, their elasticity, as in the case of metals, was shown, with, however, apparent exceptions in the case of potassium iodid, bromid, and chlorid, which seem to pass into denser allotropic forms.

§ 129. The various results thus obtained by Spring may be included under two heads ; first, Mechanical : the welding or soldering together of the particles of bodies generally regarded as solid, causing them to become more or less completely liquid, and permitting them to assume a crystalline structure; second, Chemical : effecting (1) homogeneous integration, or the genesis of allotropic species by condensation, (2) heterogeneous integration, or the genesis of new compound species by the union of unlike species; followed, in some cases, by heterogeneous disintegration or double decomposition. It should be added that in many instances the pressures employed were inadequate to produce mechanical union or consolidation, as in the cases of amorphous carbon, graphite, silica, alumina, and calcium carbonate.

From the effect of pressure in favoring homogeneous integration, Spring concludes that, below a given temperature, "matter takes the allotropic state corresponding to the volume which it is compelled to occupy."[1] In other words, if by mechanical means we can force a homogeneous mass of a given species of matter to occupy a smaller space than hitherto, it must assume a new form, which we distinguish from the previous one by calling it allotropic. In the words of our author, "Solids behave under pressure like liquids or like gases. Matter can only be condensed by pressure in the case where it admits of an allotropic state denser than that which it possesses at the moment of

[1] "La matière prend l'état allotropique correspondant au volume qu'on l'oblige d'occuper." Spring, Sur l'élasticité parfaite des corps solides, etc.

compression." "The specific gravity of a body is characteristic of the state in which it is found at a given temperature."[1] Compare with this the language of Henry Wurtz (§ 77).

§ 130. Since, then, solids, under pressure, become adapted to the containing vessel, and even flow like liquids, and undergo a process of diffusion among each other, we are led to admit that the difference between solids and liquids, as had already been shown by Andrews in the case of liquids and gases, is one of gradation. The temporary compressibility, under great mechanical force, exhibited by solids which are not known to us in permanent allotropic forms, suggests that their momentary diminution of volume, under such conditions, may be due to a conversion into denser allotropic modifications, which, at ordinary temperatures, at least, can only exist under great pressure, and assume their original form and density so soon as the

[1] "Les solides se comportent sous pression comme les liquides ou les gas. La matière ne peut être condensée par la pression que si elle admet un état allotropique plus dense encore que celui qu'elle possède au moment de la compression." "Le poids spécifique d'un corps est caractéristique de l'état sous lequel il se trouve à une température donnée." — Spring, Sur l'élasticité parfaite des corps solides, etc.

great pressure is removed. The temporary production of such condensed and liquid species might serve to explain the process of crystallization, which, according to Spring, is effected in certain powdered metals, oxyds, and sulphids, after they have been submitted to great pressure, — precisely as ice below 0° is by compression converted into a denser liquid form, which re-crystallizes when the pressure is removed. The liquefaction observed in certain cases is not, however, as Hallock [1] seemed to expect, a general result, and indeed is only to be looked for when, as in the example just mentioned, the solid is capable of assuming a denser allotropic and liquid form. In his experiments to demonstrate that, as theory demands, there is no considerable elevation of temperature from such compression, Spring has shown that solids like azoxybenzol, melting at 36°, and even phorone, melting at 28°, give no evidence whatever of liquefaction under great pressure.

Temperature being constant, increase of pressure brings about, in certain species, a more or

[1] Amer. Journal of Science, 1887, xxxiv. 277.

less permanent change of state, marked by diminution of volume, which is chemical; and this is true alike of gaseous and of solid species. Is the statement made by Spring, that the condensation of a given species of matter by pressure is possible only when this species is capable of passing into a denser allotropic form, to be applied to the passing diminution in volume which is effected by pressure in perfectly elastic gases and solids? Do the changes thus produced indicate equally temporary changes of state, giving rise to new forms, which are so unstable that they can only exist at the temperature of experiment under unusual pressures; or must we not rather admit an alteration of volume by variations of pressure as well as of temperature, which, in either case, is independent of chemical change, but, when carried sufficiently far, permits the disturbing force of chemism to assert itself, and to bring about a more or less permanent change of state?

The dynamic action of pressure produces in gases a condensation, following Boyle's law, until the internal chemic force of integration comes into play, producing dense vapors, which pass, by insensible gradations, into liquid and solid species. On the contrary, diminished pressure, still acting according to Boyle's law, causes dynamic expansion, as in what may be called the normal action of heat, until the point is reached at which chemical disintegration, or dissociation, comes into play; a process which, at constant temperature, is, like evaporation, inversely as the pressure. In the case of diminished pressure, as in that of increasing temperature, we find points at which chemical disintegration takes the place of dynamic expansion; as is seen in the passage of hexad into dyad sulphur-vapor, and of dyad into monad iodine-vapor. We thus recognize, for gases and vapors, dynamic changes of volume, under the influence alike of temperature and of pressure, which are to be distinguished from internal changes due to chemical metamorphosis. The statement that "the phenomenon of elasticity in gases and vapors is apparently a manifestation of chemical change, or metamorphosis, giving rise to new and unstable species" (**19**), was penned with reference to such internal changes as these; that is to say, to the production of gaseous integers of different degrees of condensation, and to the variations of tension due to the formation and destruction of such species by changes of temperature and pressure.

§ 131. While the chemical changes resulting from variations of pressure are such as might be expected from the consequent changes of volume, those provoked by alterations of temperature present certain apparent anomalies which require notice. Thus, increase of temperature converts solid into liquid sulphur, and this again successively into hexad and dyad sulphur-vapors. These changes to less and less dense species are reversed by diminution of temperature, which also converts water-vapor to the denser liquid form; but a farther cooling of this changes it into the lighter form of ice. Parallel cases to this are found in the expansion of bismuth on consolidation by cooling, and in the effect of great cold in converting the denser into the lighter form of metallic tin; in all of which, as in ice, augmentation of temperature brings back the denser form; or, in other words, causes contraction, like that which takes place when silver iodid is heated from $0°$ to

116°, or Rose's fusible alloy is raised from 59° to its melting-point. Farther illustrations of the effect of elevation of temperature, within a certain range, to bring about integration instead of disintegration, are seen in the transformation thereby of white phosphorus into the denser red forms, and of amorphous into crystalline arsenic.[1] In all of these cases, however, a farther augmentation of temperature reverses the process, converting the crystalline phospho-

[1] The yellow vapor of arsenic condenses into the black amorphous species, specific gravity 4.71, identical with that got in the wet way. This begins to sublime *in vacuo* at 260°, and in an inert gas at 280°–310°; but the process becomes arrested from transformation into the gray crystalline species, specific gravity 5.71, which does not sublime, even *in vacuo*, below 360°. (Engel, Comptes Rendus de l'Acad. des Sciences, xcvi. 497, 1314.) This same amorphous arsenic undergoes a similar change by pressure, at the ordinary temperature; Spring having found that, at 6000 atmospheres, it becomes crystalline, and assumes a specific gravity of 4.91, indicating a conversion of about one-fourth part into the denser species. Other cases of the genesis of more stable allotropic species by the action of heat are seen in the change of phosphorus pentoxyd into less volatile and less soluble forms; and in the similar modifications of silicon phosphate ($SiO_2 \cdot P_2O_5$), which passes successively between 260° and 1000° into less soluble species, four in number, distinguished by different crystalline forms. (Hautefeuille and Perry, and Hautefeuille and Margottet, Comptes Rendus, xcix. 32, 789.)

rus and arsenic into tetrad vapors, and these, apparently, into dyads, changing liquid water into steam, and, at a higher point, effecting its heterogeneous disintegration. It would seem that a certain degree of thermic expansion, by disturbing the existing chemical equilibrium, permits the manifestation of the integrating tendency, which, at a still higher temperature, is replaced by the contrary tendency to disintegration.

§ 132. In considering this disintegrating action, it was suggested, in § 57, that "the power of flame or of the electric spark to effect the sudden union of chlorine and of oxygen with hydrogen may be due to the effect of intense heat in separating momentarily into simpler forms portions of these gases ; so that we may have, in the process of their combination, a union of unknown elements and less dense species." Since the publication of the first edition of this volume, there has appeared the Bakerian lecture before the Royal Society of London, by J. J. Thomson of Cambridge, On the Dissociation of Some Gases by the Electric Spark.[1]

[1] Chemical News, June 5, 1887.

Therein it is shown that, by the passage of an electric spark, three inches long, through iodine-vapor at $214°$, the pressure rapidly increased, and finally became steady, remaining so for some hours, the vapor being reduced nearly one-half in density, and becoming somewhat lighter in color; so that, as remarked by the observer, the dissociation of iodine-vapor, at $214°$, by the electric spark, is as great as that got by Victor Meyer at a temperature of $1500°$. The silent discharge produced a similar effect. With bromine-vapor, the passage of the spark caused a considerable increase of pressure, which, however, disappeared almost immediately, from which he concludes a dissociation, followed by a rapid redintegration.

If now we ask at what point a simple reduction of pressure, at ordinary temperatures, may lead to chemical disintegration, our only guide is, as yet, the remarkable spectroscopic studies of Crookes, made in glass vessels, with air, hydrogen, or carbon dioxyd, exhausted to a few millionths, or even to less than one-millionth of an atmosphere, and offering what he has called an "ultra-gaseous state of matter," which opens, as he has well said, a new world to chemistry.[1]

§ 133. The question of dissociation in stellar worlds has been discussed at some length in §§ 15-20, where it was shown that the writer had already, in 1874, and again in 1881, published the suggestion that the element in the solar corona giving the line 1474 K may represent the most attenuated form of matter known to us; while later, in 1886, Crookes suggested that it might be rather the hypothetical helium of Frankland, giving the line $D_3$; regarding all which it was said by the writer that "it may be a question to which the primacy should be conceded." More recently, Grünewald of Prague,[2] from a study of the spectrum of hydrogen, concludes that this gas (which he supposes to be dissociated in the sun) is made up of four volumes of an element, the lightest of all,

[1] L., E., and D. Philos. Magazine, 1878 (5), vii. 57-64, and, farther, On the Spectroscopy of Radiant Matter, Proc. Royal Society of London, 1881, xxxii., and Philos. Transactions, 1883, clxxiv. 891-918.

[2] On the Spectra of Hydrogen, Oxygen, and Water, Astron. Nachrichten, 2797, and L., E., and D. Philos. Magazine, October, 1887.

which probably enters largely into the solar corona, and is hence called by him coronium — giving the line 1474 K — and one volume of a denser element, helium, to which belongs the line $D_3$.

§ 134. When we accept the conclusion announced in 1853, that "the doctrine of chemical equivalents is that of the equivalency of volumes" (4); or, in other words, that the law of volumes, hitherto applied only to gases and vapors, extends alike to liquid and solid species, for all of which, under proper conditions of pressure, the volume is a constant quantity, we are prepared to understand that hydrogen gas, which, as the lightest known species, is the unit of integral weight in chemistry, may also with advantage be made the unit of specific gravity for all species whatsoever, as has already been indicated in § 68. The weight of a given volume, as, for example, a litre, of any species of gas or vapor, as compared with that of the same volume of hydrogen gas at the same temperature and pressure, is the equivalent weight of that species, and "is at the same time its specific gravity, hydrogen being unity" (§ 94). Passing thence to a liquid species, as water, the weight of a litre of which at 100° is known (and is not sensibly affected by slight variations of pressure), we find that this volume of water at the temperature named has the weight of 10,703.3 litres of hydrogen at 0° and 760 mm., or of 1191.9 litres of water-vapor at the same temperature and pressure, having an equivalent weight of 17.96 ($H_2 = 2$). Neglecting the fraction, as we have before done (§ 46), water is thus represented by 1192($H_2O$), making its equivalent weight 21,408, hydrogen being unity. The law of volumes being universal, we compare the weight of water at 100° with that of the same volume of hydrogen at the standard temperature and pressure of 0° and 760 mm., and thus determine its equivalent or integral weight, from which may be calculated that of any liquid or solid whose specific gravity is known.

§ 135. As was pointed out by the writer in 1853,[1] to determine the degree of "poly-

[1] In connection with the writer's discussion of polymerism in 1848 (§ 5), it may be said that the essay by Louis Henry on "The Polymerization of Metallic Oxyds," which is noticed and cited in §§ 74–76 from an English translation published in

merism" or of "condensation" in liquid species, we must take "the specific gravity at their boiling-points" (§ 21), at standard pressure. The weight of the given volume under these conditions gives us the specific gravity on the hydrogen basis, while for any non-volatile solid species we take "the weight of the same volume at the highest temperature which that species can sustain without undergoing a change of state. In other words, the weight of liquids should be compared with that of the gaseous unit of hydrogen at the temperature at which they are generated by the metamorphosis, through condensation, of the corresponding gaseous species. In like manner, the weight of solids should theoretically be determined at the temperature at which they are generated by the metamorphosis of gaseous or liquid species. The weights thus obtained for equal volumes of the various liquid and solid species, as well as for the gaseous species, are evidently the specific

the Philosophical Magazine for August, 1885, first appeared in 1879 as "Études de Chimie Moléculaire, 1<sup>re</sup> partie; les Oxides Métalliques," in the Annales de la Société Scientifique de Bruxelles.

gravities of these species, that of hydrogen being unity ($H_2 = 2$). They are at the same time the integral weights of the species compared."

"The advantages of this natural unit of specific gravity over the arbitrary ones hitherto adopted are evident. That which, for motives of convenience, has hitherto been employed for gases and vapors, namely the weight of dry atmospheric air at standard temperature and pressure, is the weight of a mixture of gases and not of a chemical integer; while water, adopted as the unit for liquids and for solids, is, on the contrary, a liquid integer, condensed by cooling from 100°. (the temperature of its production at 760 mm. pressure) either to 15° or to 4°, its maximum density. The specific weight of hydrogen at standard temperature and pressure is the truly scientific unit of specific gravity for all bodies, whether gaseous, liquid, or solid; the integral weight of this unit entering as a factor into the specific gravities thus calculated, and causing these to represent the equivalent or integral weights of the species compared."

"It is obvious that the integral weight of water must in like manner enter into the specific gravities of liquids and solids for which this species is assumed as unity." "But this integral weight is that of a body a litre of which, though weighing 1000 grammes at 4°, weighs but 958.78 grammes at 100°; so that, in calculating specific gravities by comparing the weight of bodies with that of an equal volume of water, it is with water at the lesser density that the comparison should be made. Since most liquid and solid species expand, like water, by heat, it is theoretically desirable to take the volume of these at the highest temperature which they can sustain without change of state. This point in many cases is hard to fix. The anomalous changes in density of solids like silver iodid, Rose's fusible metal, and metallic tin have already been noted in § 131. These, not less than the ready transformation, by expansion, of aragonite into calcite by heat, and the fact that quartz and many native silicates, at temperatures below their melting-points, acquire that increased solubility which they possess (with augmented volume) after fusion, — all tend to show that chemical changes hitherto unsuspected may take place by heat in other solid species. They show, moreover, the uncertainty which attends any attempt to fix by calculation the augmentation of volume in a solid species by heat beyond the limits within which experiment has determined the rate of expansion."

§ 136. "It is clear that if such a non-volatile species could, without chemical change, be heated to a point at which its augmentation of volume from $4°$ would be just equal to that of water from $4°$ to its boiling-point at 760 mm., its specific gravity, as compared with this liquid at $100°$, would be the same as that found for the same species at $4°$ compared with water at the same temperature. The cubical expansion of water for each degree between $4°$ and $100°$ is approximately .00043, while that for iron from $0°$ to $300°$ is given as .000044, and that for quartz from $0°$ to $100°$ as .000040. The coefficients of expansion for these bodies being thus about one-tenth that of water, it would be necessary, in order to attain an augmentation comparable to that of water at 100°, to raise them to 1000° and upward, or to temperatures much above those required to produce chemical changes in quartz and in most native silicates. The coefficient of cubical expansion of these last for each degree between 0° and 100° is, moreover, much less than that for iron and for quartz, varying in general from .000020 to .000028. The latter rate of expansion, if constant to 500° (even below which many species are chemically changed), would reduce the specific gravity of the species at that temperature only 0.014; and up to 1000° only 0.028. The differences between these numbers and 0.043, which represents the cubical expansion of water from 4° to 100°, thus mark the extent of the errors involved in the determination of the specific gravities of such species with water at 4°. The imperfections and impurities of most natural and artificial crystalline species introduce errors not less considerable in the determinations of their specific gravity. The best figures obtained for such species involve, in most cases, possible deviations, in the one or the other direction, as great as those due to the different rates of expansion of water and these species; so that we may take the approximative specific gravities got with water at 4° (or better at 15°) as an available basis for fixing the integral weights of most solid species." (**19.**) (*Ante* § 69).

§ 137. We calculate the integral weights of liquid and solid species by comparing their densities with that of water. Calcite, for example, is formed by the condensation of an unknown number of volumes of calcium carbonate, $CCaO_3 = 100$. But the integral weight of water being in round numbers 21,400 (§ 46), that of calcite, sp. gr. 2.730, is 58,422, dividing which by the integral weight of $CCaO_3$, we find the number of volumes of this hypothetical carbonate condensed in one volume of such calcite, — or, in other words, its coefficient of condensation, — to be 584.22. Rejecting the fraction, we have calcite, $584 (CCaO_3) = 58,400$, with a specific gravity of 2.729, which is about that of the most common variety of calcareous spar (§ 65).

The relation between the integral weight of the calcium carbonate (hydrogen $= 1.0$ being unity) and the specific gravity of the calcite (water $= 21{,}400$ being unity) is got by the equation,

$$2.729 : 100 :: 1 : x = p \div d = v = 36.643.$$

This, the so-called molecular or atomic volume of $CCaO_3$, as it exists in calcite of such density, is the reciprocal of the above coefficient of condensation, since $36.643 \times 584 = 21{,}399.5$. But the calcite of this density is 58,400 times heavier than the same volume of hydrogen at $0°$ and 760 mm., so that, substituting this value (its specific gravity on the hydrogen basis) in the above equation, we find the value of $v$ ($H = 1.0$) to be .0017123.

$$1 : 21{,}400 :: .0017123 : 36.643.$$

In like manner, water, taking itself as the unit of specific gravity, gives $v = 17.96$; while on the hydrogen basis, with an integral weight of 21,400, we find $v = .000839$. But $1192 \times .000839 = 1.00088$.

§ 138. "In inquiries into the so-called molecular or atomic volume of liquid and solid species from the time of Leroyer and Dumas, while using the same symbols, and making $p$, as above, to represent the specific gravity on the hydrogen basis, $d$ has been taken on the basis of water as unity ($1 = 21400$); so that, having employed the ordinary formula for molecular volume, $p \div d = v$, we multiply the value of $v$ thus obtained by the coefficient of condensation and get, not 1, but 21400. Otherwise, dividing the value of $v$ by 21400, we obtain the reciprocal of the coefficient, as before, upon the hydrogen basis. Chemical integration being effected, not by juxtaposition of molecules, but by identification of volumes, the so-called molecular volume of a given liquid or solid species is thus the reciprocal of its coefficient of condensation." (**19.**) (*Ante* §§ 61–69.)

§ 139. For the better comparison of the value of $v$ in such species, the writer, in his studies of silicates, oxyds, and other non-volatile species, has used a simple device to fix an arbitrary unit for $p$. The formulas of many silicates present, in the ordinary notation, very various degrees of complexity. We find, moreover, in such species, the divalent oxygen partially replaced by monovalent fluorine or chlorine, as in chondrodite, topaz, scapolites, sodalite, and pyrosmalite; or by sulphur, as in helvite and danalite.

For all such compounds, "we assume as the unit for $p$ a weight including that of $H = 1.0$, of $Cl = 35.5$, or of $O \div 2 = 8.0$. By thus adopting a combining-weight of 8.0 for oxygen as a basis, we get a unit which gives a common term of comparison for oxyds, sulphids, chlorids, fluorids, and for intermediate compounds like the oxysulphids and oxyfluorids common in native species. It is, of course, a hypothetical unit, which, for elemental species, and for fluorids, chlorids, etc., corresponds with the normal vaporous species; but for oxydized species is some fraction thereof, as in the cases of water-vapor, $H_2O$, of spinels, and other oxyds.

"We may readily extend this system of hypothetical units from silicates to carbonates, sulphates, phosphates, and more complex species, by dividing in all cases the empirical equivalent weight by twice the number of oxygen portions ($O = 16.0$) plus the number of chlorine or fluorine portions. We have for example:—

|  |  | $p$ |
|---|---|---|
| Forsterite | $SiMg_2O_4 = 140 \div 8$ | 17.50 |
| Calcite | $CCaO_3 = 100 \div 6$ | 16.66 |
| Karstenite | $SCaO_4 = 136 \div 8$ | 17.00 |
| Gypsum | $SCaO_4.2(H_2O) = 172 \div 12$ | 14.33 |
| Apatite | $3(P_2Ca_3O_8).CaF_2 = 908 \div 50$ | 18.16 |

"In the writer's late essay on A Natural System in Mineralogy,[1] the values of $p$ have been thus determined. These silicates are there represented by a new notation, which employs symbols in small letters to represent quantivalent ratios; the combining-weights of the elements being divided by their valency, and in all cases followed by their coefficients. The formula of forsterite thus becomes $(mg_1si_1)o_2$, that of orthoclase $(k_1al_3si_{12})o_{16}$, and that of topaz $(al_3si_2)o_4f_1$." (**18**.)

While a similar unit is equally applicable to all haloid species, it has been found more convenient for metalline species, including unoxydized metals and their compounds with one another and with arsenic, antimony, sulphur, selenium, and tellurium, to divide the formula by the sum of the valencies therein represented; so that for all such the unit $p$ gives, not the mean integral weight of an oxygen compound in which $O = 8$, but that of the element, corresponding to $S = 16$, to $Fe = 28$, and to $Ag = 108$.

§ 140. "Such fractional units are convenient for the purpose of comparing the varying con-

[1] Mineral Physiology and Physiography, 279–401.

densation in species belonging to the respective groups; but it will be borne in mind that, in order to construct formulas which shall represent the true equivalent weights of liquid and solid species, we must multiply and not divide the formulas hitherto accepted as representing the normal species. The combining-weight of these must be the unit for fixing the equivalent weights and the true formulas of such liquid and solid species; which are generated by integration or polymerization from the normal species. This, though known to us in the volatile elements, and in compounds like carbon dioxyd, water-vapor, formic and acetic aldehydes, and pentine, is unknown in the case of bodies like carbon, silicon, silicon dioxyd, and most of the solid oxyds; as also in the various silicates, carbonates, sulphates, and phosphates. For all these we choose, as representing the normal species, the simplest formula which satisfies the relations of valency, and which corresponds to the theoretical gaseous or volatile species." (**18.**)

We have thus endeavored, within the limits of this supplementary chapter, to resume the principal points in the four papers named in § 99, and, in so doing, to contribute some material to the construction of what we believe will be the chemistry of the future.

# APPENDIX.

*Hardness and Chemical Indifference.* The conclusion was announced by the writer in 1863 and 1867 (§§ 28, 29), from his studies of many native mineral species, that for related solids hardness and chemical indifference increase with the condensation; or, in other words, vary *inversely* as the empirical equivalent or so-called atomic volume, represented by $v$.

$$\frac{atom.\ wt.}{sp.\ gr.} = atom.\ vol. = \frac{p}{d} = v.$$

This is in accordance with earlier observations as to the tenacity and hardness of metals. Wertheim, from his experiments, concluded in 1841 (as Guyton Morveau had long before) that the order of the common metals for tenacity and hardness is practically the same, and that tenacity varies *directly* as the quotient of the specific gravity divided by the atomic weight. This was confirmed by the studies of Calvert and Johnson in 1859, on the hardness of metals and alloys, by those of Karmarsh on the tenacity of metallic wires, and those of Bottone in 1873, also on the hardness of metals; the latter maintaining that for homogeneous bodies of uniform texture hardness and tenacity vary together, in accordance with the law enunciated by Wertheim. This law thus arrived at by various investigators for metals and alloys, as regards hardness and tenacity, is identical with that independently observed by the writer, not only for the hardness, but for the indifference to chemical action, of related silicates, carbonates, and oxyds. (See The Hardness of Metals, by Thomas Turner; Chemical News, April and May, 1887.)

*Integral Weight of Water.* In the first edition of this volume, as well as in previous notes in 1886, the weights of water-vapor and of hydrogen at 100° and 760 mm. were compared with that of water at 100°, and the ratio 1 : 1628, which represents the relative volumes of equal weights of steam and water under these conditions, was by inadvertence made to represent the condensation from which to deduce the formula and the integral weight of water; thus made $1628(H_2O) = 29,244$, instead of $1192(H_2O) = 21,408$ (§ 46). Nevertheless, in § 95 the density of "hydrogen at 0°" was already insisted upon, and later, in a paper on Chemical Integration (Amer. Jour. Science, August, 1887), it was said, instead of "the weight of atmospheric air at 0° and 760 mm. . . . as the unit of specific gravity for all gases and vapors, hydrogen gas at the same temperature and pressure, which gives us the unit of integral weight, would, however, seem to be the natural unit of specific gravity for all bodies whatsoever"; and again, that, in comparing the densities of liquid and solid species "with the density of the hydrogen unit, $H_2$ . . . we get the specific gravity of these bodies, the diad integer of hydrogen at 0° and 760 mm. being unity." In a farther paper on Integral Weights (L., E., and D. Philos. Mag., October, 1887), it was repeated that, a litre of hydrogen "at 0° and 760 mm. being assumed as a unit of volume for all species, the weight of a litre of any other gas or vapor at the standard temperature and pressure is its integral weight. In like manner the integral weight of a liquid species is the weight of the same volume at its boiling-point, under a pressure of 760 mm." (see farther § 135). Notwithstanding these repeated statements, the erroneous value for the integral weight of water, calculated from hydrogen at 100° instead of 0°, was by an oversight retained until corrected by the author in a note of The Integral Weight of Water, in the L., E., and D. Philos. Mag. for April, and also in the American Journal of Science for May, 1888.

# INDEX.

The black-faced figures in parenthesis, — thus (1), — alike in the Index and in the text, correspond to those prefixed in § 1 and § 99 to the titles there numbered, which are printed below in capitals.

Acetylene, formation of, 139
Acids, inorganic, complex, 49 ; fatty, vapor-density of, 122, *note*
Action by presence, 14, 141
Alcohol, butylic, isomers of, 91, 145; vinic, diffusion of, 132
Alcohol, critical point of, 186, 190
Aldehyde, acetic, 83; polymers of, 89, 105, 141 ; formic, 84; polymer of, *ibid.*
Allomerism, 46, 158
Allotropism as polymerism, 11, 177, 202, 205-207, 212
Alloys, fusible, 212 ; formed by pressure, 201, 202
Alumina, unit-weight of, 82 ; density of, 119
Ammonias, compound, 157
Amphibole, 47, 57, 58
Andrews, Thomas, on carbon dioxyd, 179, 184; on dense vapors, 90, 183 *et seq.*; chemical change, 193; gaseous state of matter, 182; continuity of gaseous and liquid states, 184 *et seq.*
Aniline compounds, their equivalent weights, 48
Anisomeric homologues, 45
ANOMALIES IN ATOMIC VOLUME, ETC. (1), noted, 1 ; cited, 11
Antimony, vapor of, 173; alloys with zinc, 46, 158
Aragonite, 53, 93, 220 ; density of, 105 ; formula of, *ibid.*
Armstrong and Tilden on turpentine-oils, 87, *note*
Arsenic, allotropic changes of, by pressure, 202, 212; by heat, 212, *note*
Arsenids of metals, quantivalent ratios of, 162; formed by pressure, 201

Arsenous oxyd, 122, *note*
ATMOSPHERE, THE CHEMICAL AND GEOLOGICAL RELATIONS OF (13), noted, 4; cited, 34
Atmosphere, part of the interstellar medium, 33
Atomic hypothesis, history of, 60 *et seq.*; Dalton on, 60, 62, 67; Brodie on, 67; Wright, C. R. A., on, 67, *note;* Stallo on, 62, *note;* Whewell on, 61, *note;* Cooke, J. P., on, 169; in chemistry, 62, 65, 66, 115, 118, 123, 150, 155, 169, 192, 194, 225; volumes, *see* Molecular volumes
Atomicity, hypothesis of, considered, 162 *et seq.*

**Benzoic acid**, its solubility, 116
Bernoulli, kinetic theory of gases, 61
Biotics defined, 18, 19; relations of, 21, 152
Boiling-point of liquids, 40, 218
Boutlerow on definite proportions, 46, 159, 166
Boyle's law, 183, 191, 209
Breithaupt, A., porodic bodies, 54, *note*, 132; carbon-spars, 103
Brodie, Benjamin, Calculus, 26, 27; Ideal Chemistry, 29, 31; ideal elements, 26, 29, 30; atomic hypothesis, 67
Bromine, its change by heat, 94, 122, *note;* by electricity, 214
Bunsen, fusion under pressure, 197
Butane, its equivalent weight, 110
Butylic alcohol, isomers, 91, 145

**Cagniard de Latour** on dense vapors, 90, 178–181, 190
Cailletat, chemical effects of pressure, 199
Calcite, supposed molecular structure, 66; densities, 53, 101, 104 *et seq.;* species of, 104; formulas of, 106, 223
Calcium carbonate, theoretical, integral weight of, 224
Camphenes, 80, 96
Carbon, chemistry of, 44; series, 44, 49; polymers of, 12, 42, 141, 142; carbon dioxyd, polymers of, 97, 107, 142; spars, *see* Calcite.
Carbon, crystallization of, as diamond, 189
Carbon dioxyd, liquid, 179, 184–186
Carbonites, the genus, 103
Catalysis defined, 14, 141
CELESTIAL CHEMISTRY FROM THE TIME OF NEWTON (15), noted, 4; cited, 26–35
Changes of state, chemical, 75, 135, 136, 144; defined, 196
Charcoal, an anhydrid, 12

## Index. 235

Chemical, activities, confounded with dynamics, 19-21, 65, 150, 152, 193, 195; distinguished therefrom, 209 *et seq.*; change related to pressure, 151; and mechanical forces correlated, 197, 198; elements, origin of, 9, 23 *et seq.*, 37, 43, 94; indifference related to equivalent weight, 51 *et seq.*, 59, 88, 99, 118 *et seq.*, 145; molecules, 63, 114; process, nature of, 7 *et seq.*, 22, 80, 139; species, integral, 15, 17, 22, 62, 66, 150
CHEMICAL CHANGES, ON THE THEORY OF, ETC. (4), noted, 2; cited, 7-10, 15, 22, 38-40, 44, 62, 66, 68, 70, 73, 216
CHEMICAL CLASSIFICATION, ON SOME PRINCIPLES TO BE CONSIDERED IN (2), noted, 1; cited, 11, 12; REMARKS ON (1). *See* ANOMALIES IN ATOMIC VOLUME, ETC.
CHEMICAL HOMOLOGIES, ILLUSTRATIONS OF (6), noted, 3; cited, 47, 106
CHEMICAL INTEGRATION (18), noted, 153; cited, 155-158, 173, 174, 196, 226-228
CHEMISM AS RELATED TO TEMPERATURE AND PRESSURE, noted, 154
CHEMISTRY, CELESTIAL, FROM THE TIME OF NEWTON (15). *See* CELESTIAL CHEMISTRY, ETC.
OF PRIMEVAL EARTH (11), noted, 4; cited, 26, 28
LAW OF VOLUMES IN (17), noted, 5; cited, 62-64, 78
THEORETICAL, A CENTURY'S PROGRESS IN (12), noted, 4; cited, 17, 22, 24-26, 31-33
relation of, to dynamics, 20, 21, 65, 152, 195, 209 *et seq.*; to biotics, 21, 152
Chloral, its polymerization, 84
Chlorhydric acid, genesis of, 9, 176; H. Wurtz on its volume, 124
Chlorids in aqueous solution, 22; as normal species, 119
Chlorine, its change by heat, 94, 122, *note*
Chrysolite, 56, 58
Clarke, F. W., celestial chemistry, 31, 35
Clifford, W. K., chemistry and dynamics, 20
Cobaltic ammonia salts, 48, 79
Coefficient of condensation, 64, 108, 120, 148, 223
Cold, action of, on tin, 92, 202, 211; on iron, 93; relations to chemical change, 211
Colloids, their nature, 116, 132; identical with porodic bodies, 132; their significance in mineralogy, 54; action of fluorhydric acid on, 58, 59
Colophene, 86-88, 96

Combination, chemical, Deville on, 136, 195
Complex inorganic acids of Gibbs, 49
Condensation, related to hardness, 51, 52 *et seq.*, 88, 97, 118, 145; to chemical indifference, 52, 59, 88, 145; in liquid species, 40; heat evolved in, 87, 88
Continuity of liquid and gaseous states, 184, 185
Cooke, J. P., on equivalent weights, 26; allomerism, 46, 158; definite proportions, 166, 168; atomic hypothesis, 169, 170
Copper mattes, dissociation in, 164; composition of, *ibid.*, *note*
Corona, the solar, 32, 37, 143, 215
Coronium, a supposed element, 216
Cosmogony, 23, *note*, 24
Cowles's electric furnace, 97, *note*
Crafts, J. M., action of heat on iodine, 94
Critical point of bodies, 185, 186, 195; of solutions, 187; of ether, 186; of water, 188; of alcohol, 187–190
Crookes, W., genesis of elements, 36, 37; radiant matter, 214; elemental matter, 215
Cryohydrates, Guthrie on, 115
Crystalline individuality, 17, 18, 39; form as related to volume, 71, 73 *et seq.*, 146
Crystallization from solutions, 140; under pressure, 204, 208
Cyanite, 52

**Dalton**, J., atomic hypothesis adopted by, 61, 62, 67, 68
Datolite, chemical indifference of, 57
Daubeny, C., the atomic hypothesis, 61, *note*, 69, *note*
Decomposition, double, 9, 13, 14, 98, 140, 176, 206
Definite proportions, doctrine of, questioned, 46, 157, 166, 168; defended, 170
Deliquescence, 133
Dense vapors considered, 90, 178 *et seq.*, 191 *et seq.*
Density, of solids, relation to equivalent weight, 12, 38, 39, 41, 47–49; a function thereof, 118, 120, 149; of water, steam, and hydrogen, 76–78, 147; change of, its significance, 125; variations of, in related solids, 102; relation of, to temperature, 110; law of, 100, 103; of vapors of elements, 173, 210 *et seq.*; units of, for gases and solids, 107, 108, 146
Depolymerization, 81, 141
Deville. *See* Sainte-Claire Deville, H.
Diamond, production of, 189

Diethylamine, 115
Differentiation, chemical, 14, 16, 81, 86
Diffusion, liquid, Graham on, 75, 129-135; gaseous, 128, 129, 136; in solids, 201, 207; chemical relations of, 128, 135; applied to chemical analysis, 131; compared to volatilization, 128, 131; molecules of Graham, 130
Di-pentine, 87
Dipyre, 53
Disintegration, chemical, 17, 80, 175, 214
Disintegrator, heat the universal, 151
Dissociation, chemical, a universal law, 28 *et seq.*, 35, 94, 151, 175; celestial, 27, 28, 32, 143, 175, 215; by electricity, 212; compared to vaporization, 128, 134-136; Sainte-Claire Deville on, 27, 28, 138 *et seq.*, 195
Di-terpene, 86
Dobereiner on chemical triads, 24
Drion on dense vapors, 90, 178-180
Dumas, J. B., origin of elements, 24-26, 32 *et seq.*, 35; on atomic volumes, 71, 73, 224; conferences with, 73
Dynamics, relation of, to chemistry, 19 *et seq.*, 65, 150, 152, 193, 196, 209 *et seq.*

**Earth,** CHEMISTRY OF THE PRIMEVAL (11). *See* CHEMISTRY OF THE PRIMEVAL EARTH
Efflorescence, 133
Elasticity of solids, 203, 205; of gases and vapors, 210
Electricity, 95, 98, *note;* in chemical changes, 152; in dissociation, 212
Elements, chemical. *See* Chemical elements
Energy in nature, 21, 152
Epidote, 56
EQUIVALENT VOLUMES, ON, ETC. (4). *See* CHEMICAL CHANGES, ON THE THEORY OF; also, MINERAL SPECIES, ETC.
Equivalent weights, elevated, 47-49, 79, 112; of carbon-spars, 47, 101 *et seq.;* of feldspars, 47; of pyroxenes, *ibid.;* of water and ice, 78, 109, 149. *See* Density, etc.
Ethyl oxyd, vapor of, 178; chlorid, *ibid.*, 179
Euphotide and saussurite, on, 51
Eutetic bodies, 116
Eutexia, Guthrie on, 116, *note*
Evaporation, related to diffusion, 128, 131, 133, 137; a chemical process, 137, 138, 140, 144, 195

Expansion by heat, of gases, 181, 194; of liquids, 179; of dense vapors, 182; of solids, 181

**Faraday** on chemistry and dynamics, 20
Fatty acids, vapor-density of, 122, *note*
Favre and Silbermann, thermo-chemistry, 39, 113; chemical changes, 133, 137
Faye's solar hypothesis, 27
Feldspars, 47, 56, 58
Fluorhydric acid, density, and dissociation of, 177, 178, 182, 194
Fluorine, Moissan on, 177
Formulas, structural, in chemistry, 15, 45, 67, *note*, 97, 165
Fouqué, fluorhydric acid on silicates, 56
Franke on manganese oxyds, 160
Fusibility related to condensation, 119, 145; modified by pressure, 197, 208

**Ganot**, Éléments de Physique, 76
Garnet, 56
Gaseous state, continuity with liquid, 184, 185, 188, 207
Gases, densities of, related to solids, 38, 70, 107, 110, 147, 149; unit of density for, 107, 146; law of diffusion of, 129; liquefaction of, 180; perfect, defined, 183, 193; relations of, to pressure, 184; absorbed by solids, 189; kinetic theory of, 61. *See* Normal or gaseous species
Gaudin, crystalline molecules, 16
Gay-Lussac, law of volumes, 69, 155, 191; law of thermic expansion, 191
Gems, order of, 54
Genealogy, chemical, 9, 15, 97, 165
Genesis of chemical elements. *See* Chemical elements, origin of
Geogeny, 19, *note*, 23, *note*
Gerhardt, Ch., on progressive series, 43; on structural formulas, 45
Gibbs, Wolcott, on polytungstates, 48; complex inorganic acids, 49
Gibson and Morrison, on metallic oxyds, 160
Gmelin, L., on volumes, 71; condensation of oxyds, 88
Gore on fluorhydric acid, 178
Graham, Thomas, on liquid diffusion, 75, 128, 133; relations of, to volatilization, 137; colloids, 132; polymerization in solution, 129–131; conferences with, 138

## Index. 239

Guthrie, Fred., cryohydrates, 115; physical molecules, 115-118; eutexia, 118
Gypsum, 98

Haas, metallic oxyds, 160
Hallock, effect of pressure on solids, 208
Halloysite, 58
Hannay, J. B., and Hogarth, J., solubility of solids in gases, 186-188; critical points of alcohol and saline solutions, 187; formation of diamond, 189
Hannay, J. B., absorption of gases by solids, 189
Hardness, its relation to condensation, 51, 52, 88, 97, 118, 145
Hautefeuille, polymerism by heat, 212, *note*
Heat, its relation to chemical changes, 136, 151; expansion by, 211, 212, 220 *et seq.*; evolved in polymerization, 87 *et seq.*, 185, 197, *note;* as a cause of polymerization, 211, 212
Hegel on chemical change, 16, 17
Helium, a supposed solar element, 37, 216
Helmholtz on chemistry and dynamics, 20
Henry, Louis, on polymerization, 118 *et seq.*, 217, *note*
Heptine, 86
Herepath, J., kinetic theory of gases, 62
Hinrichs, G., on primary matter, 33, *note*
Hogarth, J. *See* Hannay and Hogarth.
Holoëdrites, the genus, 105
Homogeneous chemical species. *See* Chemical species, integral
Homologous series, history of, 43, 44, 49, 145; their universality, 44, 49, 145, 155, 158
Homologues, anisomeric, 45; isomeric, *ibid.*
Hopkins, Wm., on fusing-point of solids, 197
Hornblende, 52, 53
Huggins on elemental matter, 33, *note*
Hydrocarbons, quantivalent ratios of, 165
Hydrofluoric acid, action of, on silicates, 56-59
Hydrogen gas, a unit of weight, 172, 216; of specific gravity, 109, 216-219; solubility of sodium in, 188
Hylogeny, 23, *note*

Ice, its density, 78; fusing-point lowered by pressure, 198, 202, 211; relations to water, 91, 144
Identification, chemical, 14, 16, 146

Insolubility related to condensation. *See* Condensation related to chemical indifference
INTEGRAL WEIGHTS IN CHEMISTRY (19), noted, 154; cited 175, 218–225
Integration, chemical, 17, 139 *et seq.*, 143, 150, 151, 153
Interpenetration in chemistry, 15, 16
Iodine, its metamorphosis by heat, 94, 122, *note*, 143, 173, 182, 194, 210; its dissociation by electricity, 214
Iolite, 56
Iron, supposed effect of cold on, 93
Isambert, studies in dissociation, 195
Isomeric homologues, 45
Isomorphism, relation to volume, 71–73, 146; polymeric, 158
Isoprene, 86, 96

Jade, SUR LA NATURE DU (9), noted, 3; cited, 52, 53
Jadeite, 53

Kant, Em., on chemical union, 15 *et seq.*
Kinetic theory of gases, 61
Kirchhoff, G., on spectral lines, 37, 95, 215
Kopp, H., atomic volume, 72; density of water, 77

Laurent, A., numeric harmony, 155
Lavoisier on primal matter, 33, 35
Leroyer, atomic volumes, 71
Leucite, 58
Light in chemical change, 152
Limonenes, 85, 96
Liquefaction of solids under pressure, 198, 204, 205, 208
Liquids as polymers, 41, 48, 142; determination of densities of, 40; diffusion in, 129–131
Lockyer, N., solar physics, 29, 32

Mackintosh, J. B., action of fluorhydric acid on silicates, 56 *et seq.*; on an iron-nickel sulphid, 162
Macvicar, J. G., on crystalline molecules, 16
Magnesian phosphate, 88, 98
Magnetite in copper matte, 164, *note*
Mallet, J. W., on fluorhydric acid, 177

Manganese, oxyds of, 160
Marriotte's law, 181. *See* Boyle's law
Materia prima, 17, 23, 33, *note*, 36, 95, 143
Mathesis, 23, *note*
Matter, divisibility of, 60 ; continuity of liquid and gaseous states of, 182 *et seq.*, 188, 207 ; of solid and liquid states of, 207 ; radiant and ultra-gaseous, 214
Meionite, 51, 52, *note*
Mendeléef on valency, 163
Metachloral, 84
Metagenesis in chemistry, 8, 15, 45, 80, 97, 139, 143, 199
Metallates, 55
Metallometallates, 55
Metallic oxyds as polymers, 55, 88, 97, 98, 119, 121, *note*, 217, *note*
Metaldehyde, 84, 89, 121, *note*
Metals, quantivalent ratios of oxyds of, 160 ; of sulphids of, 162 ; of arsenids of, 162 ; welding of, 204
Metamorphosis in chemistry, 8, 14, 80 *et seq.*, 141, 142, 202 ; by expansion, 9, 81 ; by condensation, 10, 81
Metaterpene, 88
Methylene oxyd, 84, 121, *note*
Meyer, Victor, action of heat on chlorine, 94 ; on iodine, 214
Micas, 58
Micaceous type in mineralogy, 54
Miller, W. Allen, on dense vapors, 180, 181
Mills, E. J., genesis of elements, 36
MINERAL SPECIES, ON THE CONSTITUTION AND EQUIVALENT VOLUME OF SOME (5), noted, 2 ; cited, 45, 47, 62, 70
MINERALOGY, A NATURAL SYSTEM OF, ETC. (16), noted, 5 ; cited, 55, 57, 64, 65, 79, 80, 82, 127, 227
Mineralogy defined, 18
MINERALOGY, OBJECTS AND METHODS OF (10), noted, 3 ; cited, 16, 17, 41, 42, 47, 53, 68, 74, 117, 138
Mitscherlich, isomorphous groups, 72
Moissan on fluorine, 177
Molecular hypothesis. *See* Atomic hypothesis
Molecular volume, 63, 65, 124
Molecules, 63 ; physical, 63, 114 ; chemical, 63, 114 ; diffusion, Graham on, 130
Mono-pentine, 88
Morrison. *See* Gibson and Morrison

**Nascent state** of bodies, 98
NATURE IN THOUGHT AND LANGUAGE (14), noted, 4; cited, 14, 18, 19, 20, 21, 24, 44
Naumann, A., on dense vapors, 192, 194
Nebulæ, chemistry of, 33–35; possible origin of, 33
NEWTON, CELESTIAL CHEMISTRY FROM THE TIME OF (15), noted, 4; cited, 26–35
Newton on atomic hypothesis, 69; his theory of light, 67
Nitric tetroxyd, 122, *note*
Normal or gaseous species, 90, 91, 94, 97–99, 107, 119, 121, *note*, 142, 145, 148, 149
Notation, quantivalent, for oxyds, silicates, etc., 227
Numbers in chemistry, law of, 43 *et seq.*, 69, 145, 155; harmony of, 155

**Ocean**, part of interstellar medium, 33
Oken, Lorenz, 23; his Physiophilosophy, 23, *note*
Omnia mensura et numero, etc., 69, *note*
Ontology, 23, *note*
Opal, 53, 58
ORGANIC CHEMISTRY (3), noted, 2; cited, 44; defined, 44
Orthoclase, 58
Osmic oxyd, 121, *note*, 160
Otto on atomic volume, 71
Oxyds, anhydrous, in mineralogy, 55; solid, as polymers, 55, 88, 98, 119, 122, *note;* classification of, 121, *note;* quantivalent ratios and general formula for, 160, 161
Oxyfluorids, 226
Oxygen, its combining-weight, 77, 100, 102, 108
Oxysulphids, 226
Ozone, 11, 93, 173

**Paraldehyde**, 83, 84, 89, 122, *note*
Pentine, 86 *et seq.;* polymers of, 87, 89–91, 96
Periodic law, 46
Petalite, 56, 58
Pettenkofer, chemical elements, 24, 32
Pfaff, chemical effects of pressure, 199
Phosphate of silicon, allotropic modifications of, 212, *note*
Phosphorus, allotropic forms of, 12, 42, 91, 126, 141, 142, 151, 173, 196, *note*, 211, 212

## Index. 243

Phosphorus pentoxyd, allotropic modifications of, 212, *note*
Phylloid type in mineralogy, 54
PHYSIOLOGY, THE DOMAIN OF, ETC. (14). *See* NATURE IN THOUGHT AND LANGUAGE
Physiology, general, 19, *note;* mineral, *ibid*.
Pickering, Spencer, physical molecules, 114, 117, 131
Pinenes, 85, 96
Playfair and Wanklyn on dense vapors, 192
Pneumatogeny, 23, *note*
Polycarbonates, 47, 49, 104-106
Polymerism defined and illustrated, 10-12, 53, 114, 115, 174, 177; objection to the term, 174; in salts, 39, 40; as related to allotropism, 42, 177, 202, 206; in dense vapors, 192-194; in non-gaseous species, 41, 42, 145, 217, *note;* in solids and liquids, 48, 142; in water and ice, 40, 75, 76, 82, 91, 108, 138, 144, 147, 148; pentine, 87, 88, 89, 91, 96; aldehydes, 83, 84; of butylic alcohol, 91, 145; of various precipitates, 98; of oxyds, phosphates, and silicates by heat, 88; in solutions, Graham on, 130, 131; coefficient of, *see* Coefficient of condensation; relation to hardness, *see* Hardness in relation to condensation; to chemical indifference, *see* Chemical indifference; to specific gravity of solids, *see* Density in relation to equivalent weight
Polymolybdates, 49
Polyphosphates, 49
Polysilicates, 47, 49, 118
Polyterpenes, 86
Polytungstates, 48, 79
Polyvanadates, 49
Porodic type in mineralogy, 54, 58, 59, 132; *see* Colloids
Porodini of Breithaupt, 54
Pressure, relation to chemical change, 151, 199-208; affecting solution, 198
Priestley, J., discourse at grave of. *See* CHEMISTRY, THEORETICAL, A CENTURY'S PROGRESS IN (13)
Primal matter. *See* Materia prima
PROCESS, THE CHEMICAL, THOUGHTS ON (7), noted, 3; cited, 14, 16, 22; nature of, defined, 7-22, 139
PROGRESSIVE SERIES IN CHEMISTRY, noted, 154; 43-45, 48, 145, 155, 157 *et seq.*, 167.
Prout's hypothesis, 25, 46
Pumice, its formation from obsidian, 190

## Index.

Pyrognomic minerals, 88
Pyroxene, 47, 52, 57

Quartz, 53, 58, 107; its coefficient of expansion, 221

Ramsay, Wm., and Sidney Young on critical point of alcohol, 190; on dense vapors, 192–194
RAPPORTS ENTRE QUELQUES COMPOSÉS DIFFÉRENTS PAR $H_2$ ET $O_2$, SUR LES, noted, 156; cited, 157
Reciprocal number, of species, 110; of coefficient of condensation, 64, 108, 110, 148, 149, 224 *et seq.*
Regnault, V., weight of hydrogen, 76
Richards, R. H., metamorphosis of tin, 92
Roberts, W. Chandler, on results of W. Spring, 202

Sainte-Claire Deville, H., on dissociation, 27, 28, 133–136, 138, 195; conferences with, 137
Sajotschewsky on dense vapors and critical points, 186
Saussurite. *See* Euphotide
Scheerer, Th., on pyrognomic minerals, 88; polymeric isomorphism, 158
Schiel, James, on progressive series, 43
Schützenberger, on definite proportions, 46, 166, 167; on composition of water, 168, 171; on metallic oxyds, 159
Sciences, natural, tabular view of, 19, *note*
Sesqui-terpene, 86
Shenstone. *See* Tilden and Shenstone.
Silbermann. *See* Favre and Silbermann
SILICATES, A CLASSIFICATION OF (16). *See* MINERALOGY, A NATURAL SYSTEM OF
Silicates, tribes of, 57, 58; changes of, by heat, 222; expansion of, 222; notation for, 227
Silicon, series of, in chemistry, 44, 49; dioxyd, polymerism of, 97, 107, 119, 122, *note*, 142; unit-weight for, 82; reduction and volatility of, 98, *note*
Silicon phosphate, allotropic forms of, 212
Sillimanite, 52
Silver iodid, action of heat on, 211, 220
Silver nitrate, solubility of, 116
Sodium, solution of, in hydrogen, 188
Sodium chlorid, H. Wurtz on, 126; solutions of, 198

# Index. 245

Solids, density of, as related to gases, 38, 41, 64, 70, 81, 91, 107, 108, 119, 120, 147–149; unit of density for, 146; compressibility of, 203; elasticity of, 204, 205; as polymers, *see* Polymerism.

Solubility of oxyd compounds, 99; at high temperatures, 116; its geological relations, 117; relation of, to condensation, 145. *See* Chemical indifference, etc.

SOLUTION AND THE CHEMICAL PROCESS, THOUGHTS ON ('7), noted, 3; cited, 14, 16, 22

Solution defined as chemical, 22, 112, 142; modified by pressure, 198; as heterogeneous integration, 140; densities of Graham, 129

Solutions, gasefaction of, 187, 188

Sorby, H. C., correlation of mechanical and chemical force, 197, 198

Spar, order of, 54

Spathoid type in mineralogy, 54

Spathometallates, 55

Specific gravity. *See* Density

Specific gravity, of gases and vapors, 174; of solids, 207, 222; hydrogen gas as a common unit for, 109, 216–219; at what temperatures determined, 218, 220 *et seq.*

Spodumene, 56

Spring, W., effects of pressure on solids, 151, 199 *et seq.;* list of papers by, 200, *note;* on diffusion in solids, 201; on a manganese oxyd, 160

Stallo on atomic hypothesis, 62, *note*

Staurolite, 56

Steam, its weight, 76–78, 109; conversion into water, 75–78, 108, 109, 138–144

Stellar matter, chemistry of, 28, 31, 36, 37, 94, 215

Stoichiogeny, 23, *note*, 24, 26, 31

Stoney, Geo. Johnstone, on solar chemistry, 32

Strontian sulphate, 98

Structural formulas in chemistry, 15, 45, 67, *note*, 97, 165

Sulphids, metallic, quantivalent ratios of, 159

Sulphur dioxyd, 93, 121, *note*

Sulphur-vapor, density of, 10, 89, 96, 141, 143, 173, 182, 194, 209, 210

**Temperature**, relation to chemical change, 150, 210, 211 *et seq.*

Terpene, its liquid and solid polymeres, 85–87, 90, 91

Tetra-pentine, 88

Thilorier on carbon dioxyd, 179

Thomson, Wm., relation of pressure to fusion, 198; J. J., dissociation by electrical discharge, 213

Tilden. *See* Armstrong and Tilden
Tilden and Shenstone, solution at high temperatures, 116, 190
Tin, metamorphoses of, 92, 141, 151, 202, 211
Tourmaline, 49, 57
Trachytism, Ch. Deville on, 190
Tresca, flow of cold metals, 204
Tribes in mineralogy, 57
Tridymite, 58, 107
Tri-pentine, 87
Tschermak on crystalline admixtures, 107
Turpentine-oils, 85 *et seq.*, 89
Tyndall, J., on potency in matter, 21
TYPES IN CHEMISTRY, THEORY OF (8), noted, 3; cited, 14, 15; their significance, 15, 45, 97; in mineralogy, 54

Unit values, arbitrary, for $p$, 225, 227
Unit weights, 79, 82; unit volumes, 79, 83; of density for solids and liquids, 108, 146; for gases, 107, 146

Valency, hypothesis of, considered, 162, 163
Valerylene, 86
Vapors, dense, 90, 178–182, 191 *et seq.*; of elements, 173; defined by Andrews, 186; density of, as related to solids, *see* Solids, density of
Vital phenomena, 18
Volatilization, related to diffusion, 131, 137; compared to decomposition, 134, 195. *See* Evaporation
VOLUMES, ATOMIQUES, SUR LES, noted, 154; cited, 155, 156, *note*
VOLUMES, LAW OF, IN CHEMISTRY (17), noted, 5; cited, 62–64, 78
Volumes, significance of, 71, 79; relation to atomic hypothesis, 62, 68; to crystalline form, 71–74, 146; law of, universal, 68, 70, 71, 75, 79, 146; equivalency of, 216

Wallach on turpentine-oils, 87
Waltershausen, Von, on crystalline admixtures, 106
Water, equivalent weight of, 78, 100, 217; its conversion into steam, 75, 77, 78; heterogeneous dissociation of, 89, 133; its solvent powers at high temperatures, 116, 189; in fused rocks, 117; critical point of, 188; supposed variations in composition of, 168, 170–172
Welding of metals by pressure, 204

Whewell, W., atomic hypothesis, 61, *note*
Willemite, 58
Wright, C. R. A., on atomic hypothesis, 67, *note*
Wurtz, Ad., on polysilicates, 118
Wurtz, Henry, geometrical chemistry, 122; his scheme examined, 124-127; on significance of density, 123, 207

Zinc and antimony, alloys, 46, 158
Zircons, varying density of, 53; insolubility of, 56
Zoisite compared with meionite, 51, 52

# NEW TREATISE ON MINERALOGY.

IN PRESS.

SYSTEMATIC MINERALOGY; BASED ON A NATURAL CLASSIFICATION: BY THOMAS STERRY HUNT, M. A., LL. D., F. R. S.

The historical relations of the New System here adopted alike to the Chemical and the Natural-History methods are set forth in the essay on Mineralogical Classification in the author's MINERAL PHYSIOLOGY AND PHYSIOGRAPHY (pp. 279-401), with some modifications indicated in the Preface to the Second Edition of that volume. The treatise will include a general discussion of chemical principles in accordance with the views advanced in the NEW BASIS FOR CHEMISTRY, embracing the question of the co-efficient of condensation, the periodic law, a simplified monadic chemical notation, and the theory of compound polybasic acids of high equivalent, often very complex in constitution. Thus, for example, besides true polysilicic acids, are others including with silicic oxyd those of boron, of aluminium and of sulphur, as well as fluorine and chlorine, giving rise not only to simple silicates but to fluorosilicates, chlorosilicates, borosilicates, aluminosilicates, fluorboraluminosilicates, sulphataluminosilicates, etc. Other points will be: Arrangement of all native and artificial species of the mineral kingdom in classes and orders upon a chemical basis; Application to them of the principles of polymerism and of homologous or progressive series; Relations of the constitution of solids, not only to their specific gravity, but to their hardness and their greater or less chemical indifference, which latter are shown to be connected with the variations in condensation or so-called atomic volume; Recognition of the wide distinction between crystalline and colloid or amorphous species; Sub-division, on the above grounds, of orders into tribes genera and species, and designation of these by a binomial Latin nomenclature.

www.ingramcontent.com/pod-product-compliance
Lightning Source LLC
Chambersburg PA
CBHW032132230426
43672CB00011B/2312